THERMAL STRESSES IN A COMPOSITE CYLINDER
WITH AN ARBITRARY TEMPERATURE DISTRIBUTION ALONG ITS LENGTH

TEMPERATURNYE NAPRYAZHENIYA V SOSTAVNOM TSILINDRE
PRI PROIZVOL'NOM RASPREDELENII TEMPERATURY PO VYSOTE

ТЕМПЕРАТУРНЫЕ НАПРЯЖЕНИЯ В СОСТАВНОМ ЦИЛИНДРЕ
ПРИ ПРОИЗВОЛЬНОМ РАСПРЕДЕЛЕНИИ ТЕМПЕРАТУРЫ ПО ВЫСОТЕ

THERMAL STRESSES
IN A COMPOSITE CYLINDER

With an Arbitrary Temperature Distribution
Along Its Length

V. S. Nikishin

Authorized translation from the Russian

Springer Science+Business Media, LLC

1966

The original Russian text, published as an unnumbered «Trudy» of the Calculating Center of the Academy of Sciences of the USSR in Moscow in 1964, has been extensively revised and updated by the author.

В. С. Никишин

ТЕМПЕРАТУРНЫЕ НАПРЯЖЕНИЯ В СОСТАВНОМ ЦИЛИНДРЕ
ПРИ ПРОИЗВОЛЬНОМ РАСПРЕДЕЛЕНИИ ТЕМПЕРАТУРЫ ПО ВЫСОТЕ

Library of Congress Catalog Card No. 65-17787

ISBN 978-1-4899-5447-3   ISBN 978-1-4899-5445-9 (eBook)
DOI 10.1007/978-1-4899-5445-9

Contents

Introduction... 3

CHAPTER 1. Mathematical Formulation of the Problem..................................... 6

CHAPTER 2. The Thermally Stressed State of a Composite Cylinder with Temperature
Function $\eta_1(z)$.. 10

CHAPTER 3. Thermal Stresses in a Composite Cylinder with an Arbitrary Temperature
Distribution Along Its Height and Constant Thermal-Expansion Coefficients
in the Shell and the Core.. 29

CHAPTER 4. Thermal Stresses in a Composite Cylinder with Different Thermal-Expansion
Coefficients in the Shell and the Core and an Arbitrary Temperature
Distribution Along Its Length.. 43

CHAPTER 5. Numerical Methods for Calculating the Thermal Stresses in Composite
Cylinders with Continuous and Discontinuous Temperature Functions $T(z)$........... 44

    1. Recommendations for the Practical Use of the Tables of Stresses for the
Temperature Function $\eta_1(z)$. The Accuracy Needed in These Tables........... 44

    2. Calculation of Thermal Stresses in a Composite Cylinder with an
Arbitrary Temperature Discontinuity.. 48

    3. Calculation of Thermal Stresses in a Composite Cylinder for an
Arbitrary Temperature Function $T(z)$.. 52

    4. Calculation of Thermal Stresses in a Composite Cylinder for a Bounded
Discontinuous Temperature Function $T(z)$.. 55

Conclusion... 57

Literature Cited... 58

Appendix... 59

Introduction

Thermal Stresses in the Shell and Core of a Composite Cylinder

At present centrifugally cast cylindrical shells of reinforced concrete, 0.4 to 6 m in diameter, with wall thicknesses from 5 to 16 cm are widely used in transport and hydrotechnical construction. Thin-walled tubular shells are used in the construction of supports for bridges, viaducts, and roads, as well as piers, dams, and embankments. Centrifugally cast reinforced concrete shells of small diameter are used extensively in the construction of supports of contact grids. These shells may either be filled with concrete or may be used empty.

The high productivity and economy possible in structures using assembled cylindrical reinforced concrete members make them attractive also for industrial construction [1], and the question of the durability of cylindrical structures employing reinforced concrete shells is therefore a very timely one. Studies of many structures (bridges, roads, trestles) built on shells filled with concrete have revealed large numbers of cracks [2]. Cracks in the shells seriously affect the strength of a structure, are responsible for further breaking up of concrete, and thus reduce the working life of the structure.

One of the main reasons for the breaking of concrete shells by crack formation is the high thermal-expansion stress. Transport structures are continually subjected to the thermal action of the surrounding medium. Quite frequently, structures made of reinforced concrete shells are immersed — along their entire length — in two media at different temperatures, for instance, partially in water or soil and partially in air. If the difference between the temperature of the water or soil and that of the ambient air is large, a complex, nonuniform temperature field will be set up in the structure, with large radial and longitudinal temperature drops. Such a field leads to appreciable thermal-expansion stresses in the shell. Frequently these stresses exceed the breaking point and result in fracture of the shell.

If the thermal action of the surrounding medium on a cylindrical structure is axially symmetric, an axially symmetric temperature field will result, which in the general case will vary in both the radial and the longitudinal directions. This field may be expressed mathematically by the function

$$T = T(r, z) \qquad (0.1)$$

Below, only the axially symmetric case is considered.

If we are to prevent temperature cracks, we must first of all learn how to calculate the thermal stresses that exist in cylindrical structures due to the temperature fields found under actual operating conditions. In the general axially symmetric case, this problem reduces to the determination of thermal stresses in solid and hollow cylinders for the temperature function (0.1). The case of particular interest is that of a solid cylinder consisting of a tubular shell and a core (the material in the tube) with different elastic and thermal properties. We shall call such a cylinder a *composite cylinder* (Fig. 1). We shall assume that the shell and core materials are each — individually — homogeneous and isotropic, but that their moduli of elasticity E, Poisson ratios ν, and thermal-expansion coefficients a may differ. The total composite cylinder is therefore not homogeneous.

The problem of the thermal stresses in a composite cylinder for the temperature function (0.1) is much more complex than the same problem for a homogeneous solid cylinder. In isotropic and homogeneous structures, thermal stresses result from a nonuniform deviation of the temperature from the so-called "original temperature distribution" which exists in an elastic structure in a truly unstressed state. If the structure is not homogeneous or consists of homogeneous materials with different elastic and thermal characteristics, thermal stresses arise in it even if the temperature change from the original temperature distribution is uniform.

The concept of the original temperature distribution in concrete structures has been introduced by Prof. V. S. Luk'yanov [3]. The original temperature distribution in a concrete structure begins to form while the concrete is hardening. As the concrete hardens, the temperature distribution in the concrete structure changes continually, owing to the exothermic reaction of the cement and the thermal action of the surrounding medium. Free plastic deformation takes place in the structure until it reaches its elastic state. The temperature distribution existing at the instant when a further temperature change can no longer produce a free deformation of the structure and thermal stresses are produced in it is taken as the original temperature distribution. The problem of the determination of the original temperature distribution in concrete structures is complex and must be investigated separately. Such investigations may successfully be carried out using the hydraulic integral of V. S. Luk'yanov for the calculation of temperature fields.

Besides the problem of the determination of thermal stresses in a composite cylinder for the temperature function (0.1), it is of great independent importance, from the point of view of construction practice, to solve the special case of this problem arising when the temperature field of the composite cylinder varies only along its length (along its axis) and is given by the temperature function

$$T = T(z) \tag{0.2}$$

In this special case the temperature is constant across any cross section perpendicular to the axis of the cylinder and varies as a function of the cross section. Such a temperature distribution is observed in quite thin composite cylinders and hollow thin-walled shells immersed lengthwise in two media at different temperatures.

Small-diameter shells filled with concrete and hollow thin-walled shells of different diameters are widely used in the transport-construction industry. During the last few years a large number of footbridges, trestle bridges, and roads were constructed on supports made from centrifugally cast reinforced concrete shells with an external diameter between 0.4 and 0.6 m, filled with concrete. Now they exhibit a large number of cracks. The supports of these structures (composite cylinders) are surrounded by two different media along their height: soil and air. The temperatures of these media may be appreciably different. In such thin cylinders we can neglect the temperature variation along the radius (in transverse cross sections) in comparison with the temperature variation along the height.

The present work is devoted to the solution of the problem of thermal stresses in a composite cylinder for any temperature function (0.2), given either in analytical form or in the form of tables. The analysis pertains to a cylinder in a free (not fixed) state, in the absence of surface forces.

An exact solution of the problem is obtained here on the assumption that the axial stresses at the ends of the composite cylinder satisfy the boundary conditions in the integral sense, i.e., that they are compensated. In accordance with the Saint-Venant conditions, this assumption has no marked effect on the solution at points distant from the ends. A simple approximate method for the solution of the problem and an estimate of the error are obtained on the basis of the exact solution. It must be emphasized that the method for solving the problem used in this work also allows us to find solutions for bounded discontinuous temperature functions (0.2). This is important since temperature fields that are to all intents and purposes discontinuous may be encountered in transport-construction practice (for instance, in block laying of concrete and in the impregnation of foundations of supports of a contact grid with asphalt at a temperature of 180°C) and it is very difficult — nearly impossible — to obtain even reasonably accurate solutions by any of the known methods such as, for instance, the method of finite differences [4] or the variational method [5].

The method of solving the problem of thermal stresses in a composite cylinder for the temperature function (0.2) proposed in this paper reduces to the following. Using biharmonic displacement functions, we find the general solutions of the problem separately for the shell and for the core for the temperature function (0.2) and for an arbitrary temperature dependence of the thermal-expansion coefficients of the shell and the core. We represent the latter in the form

$$a_0 = a_0(T) = \overline{a}_0(z) \qquad a_1 = a_1(T) = \overline{a}_1(z) \tag{0.3}$$

4

We then solve the problem of the thermal stresses in a composite cylinder free of external stresses for the temperature function

$$\eta_1(z) = \begin{cases} -\tfrac{1}{2}, & -\dfrac{H}{2} \leq z < 0 \\[3mm] \tfrac{1}{2}, & 0 < z \leq \dfrac{H}{2} \end{cases} \tag{0.4}$$

for different constant values of the thermal-expansion coefficients of the shell and core materials ($a_0 = $ const, $a_1 = $ const). The solution of this problem is obtained in the form of complex functional series expressed in terms of Bessel functions of an imaginary argument of the first and second kind, and in terms of trigonometric functions. After ensuring the convergence of the series by separating out the principal parts, we calculate the stresses σ_r, σ_θ, σ_z, τ_{rz} for the temperature function $\eta_1(z)$ with η_1 η_1 η_1 η_1 the aid of a BESM-2 computer. On the basis of these calculations and using the similarity theorem, we compile tables of the stresses for cylinders of various geometrical sizes with different characteristics of the shell and core materials. The stresses σ_i ($i = r$, θ, z), τ_{rz}, obtained in the form of tables (or η_1 η_1 graphs), are then used to construct a closed solution for a thermally stressed composite cylinder free of external loads for the temperature function (0.2), for the case both of constant and of temperature-variable thermal-expansion coefficients of the shell and core materials obeying the condition

$$\gamma = \frac{a_1(T)}{a_0(T)} = \text{const} \tag{0.5}$$

The simplicity and sufficiently high accuracy of the approximate solution obtained here of the problem of the thermal stresses in a composite cylinder for the arbitrary temperature function (0.2) make it possible to use this solution in practical calculations. In Chapter 5 we present recommendations for engineers concerning the practical use of the results of this work and describe a simple numerical method for calculating thermal stresses for a given temperature distribution along the cylinder.

The problem solved here for composite cylinders had been previously solved for homogeneous solid and hollow cylinders [6]. The fact that in this work the calculations of the tables of the stresses σ_i ($i = r$, θ, z), τ_{rz} are discussed in terms of concrete composite cylinders in no way takes away from η_1 η_1 the general nature of this study. The results may be used not only in the construction field, but in all other areas of application of thermoelasticity as well.

It is hoped that our solution of the problem of thermal stresses in a composite cylinder for the special temperature distribution (0.2) will be of help in the solution of the general form of the problem, i.e., for the temperature function (0.1).

CHAPTER 1

Mathematical Formulation of the Problem

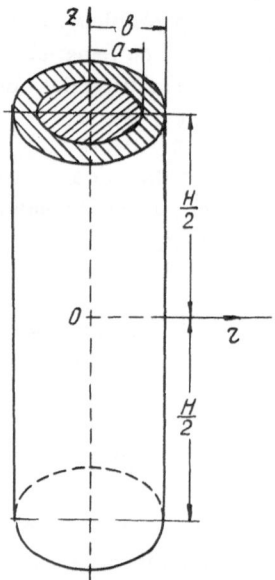

Fig. 1. Schematic drawing of the cylinder.

We consider a circular composite cylinder of length H, consisting of a shell and a core with different elastic and thermal characteristics; the internal and external radii of the shell are a and b (see Fig. 1).

The temperature field in the cylinder is assumed to be given by (0.2), and is thus a function of only the coordinate z measured along the length of the cylinder. We denote the elastic characteristics of the materials of the shell and the core by E_0, ν_0 and E_1, ν_1, respectively. The thermal-expansion coefficients will be assumed to be functions of the temperature and will be given by (0.3). We will obtain solutions for the shell and the core separately. The stresses and displacements in the shell will be denoted by σ_{r0}, $\sigma_{\theta0}$, σ_{z0}, τ_{rz0}, u_0, v_0, and w_0, and those in the core by σ_{r1}, $\sigma_{\theta1}$, σ_{z1}, τ_{rz1}, u_1, v_1, and w_1 (Fig. 2).

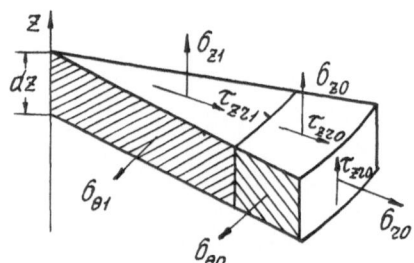

Fig. 2. An elementary sector of a cylinder.

6

The temperature function (0.2) determines a temperature field, symmetric relative to the z axis, which generates an axially symmetric stressed state. Here the stresses, deformations, and displacements in the shell and in the core are independent of the meridional angle θ, and are constant in each plane perpendicular to the z axis, which coincides with the cylinder axis. The tangential stresses $\tau_{r\theta i}$, $\tau_{\theta z i}$ and the tangential displacements v_i ($i = 0, 1$) are zero because of the symmetry.

These assumptions lead to the following thermal-equilibrium equations for the displacements in the shell and core [7]:

$$\left.\begin{aligned}
\Delta u_i - \frac{u_i}{r^2} + \frac{1}{1 - 2\nu_i} \frac{\partial e_i}{\partial r} &= 0 \\[2ex]
\Delta w_i + \frac{1}{1 - 2\nu_i} \frac{\partial e_i}{\partial z} &= \frac{2(1 + \nu_i)}{1 - 2\nu_i} \frac{d[a_i(T)T(z)]}{dz}
\end{aligned}\right\} \tag{1.1}$$

$$(i = 0, 1)$$

where $\Delta = \dfrac{\partial^2}{\partial r^2} + \dfrac{1}{r}\dfrac{\partial}{\partial r} + \dfrac{\partial^2}{\partial z^2}$ is the Laplace operator in cylindrical coordinates for the case of axial symmetry, and $e_i = \dfrac{\partial u_i}{\partial r} + \dfrac{u_i}{r} + \dfrac{\partial w_i}{\partial z}$ ($i = 0, 1$) denotes the total volume deformation of the shell and the core.

The total radial, tangential, and axial deformations are the sums of the two deformations:

$$\bar{\epsilon}_{ri} = \frac{\partial u_i}{\partial r} = \epsilon_{ri} + a_i T \qquad \bar{\epsilon}_{\theta i} = \frac{u_i}{r} = \epsilon_{\theta i} + a_i T \qquad \bar{\epsilon}_{zi} = \frac{\partial w_i}{\partial z} = \epsilon_{zi} + a_i T \tag{1.2}$$

where $a_i T (i = 0, 1)$ are the free deformations of the shell and the core due to temperature changes, and ϵ_{ri}, $\epsilon_{\theta i}$, ϵ_{zi} ($i = 0, 1$) are the deformations of the shell and core due to the thermal stresses. According to the general formulation of Hooke's law, these thermal-stress deformations are

$$\left.\begin{aligned}
\epsilon_{ri} &= \frac{1}{E_i}[\sigma_{ri} - \nu(\sigma_{\theta i} + \sigma_{zi})] \\[2ex]
\epsilon_{\theta i} &= \frac{1}{E_i}[\sigma_{\theta i} - \nu(\sigma_{ri} + \sigma_{zi})] \\[2ex]
\epsilon_{zi} &= \frac{1}{E_i}[\sigma_{zi} - \nu(\sigma_{ri} + \sigma_{\theta i})]
\end{aligned}\right\} \tag{1.3}$$

The stresses and displacements are related by the equations

$$\left.\begin{aligned}
\sigma_{ri} &= \frac{E_i}{1 + \nu_i}\left[\frac{\nu_i}{1 - 2\nu_i} e_i + \frac{\partial u_i}{\partial r}\right] - \frac{E_i}{1 - 2\nu_i} \cdot a_i T \\[2ex]
\sigma_{\theta i} &= \frac{E_i}{1 + \nu_i}\left[\frac{\nu_i}{1 - 2\nu_i} e_i + \frac{u_i}{r}\right] - \frac{E_i}{1 - 2\nu_i} a_i T \\[2ex]
\sigma_{zi} &= \frac{E_i}{1 + \nu_i}\left[\frac{\nu_i}{1 - 2\nu_i} e_i + \frac{\partial w_i}{\partial z}\right] - \frac{E_i}{1 - 2\nu_i} a_i T \\[2ex]
\tau_{rzi} &= \frac{E_i}{2(1 + \nu_i)}\left[\frac{\partial w_i}{\partial r} + \frac{\partial u_i}{\partial z}\right]
\end{aligned}\right\} \tag{1.4}$$

7

We will assume that the materials making up the shell and the core are in contact, and that there is no free axial displacement of the shell relative to the core (the two materials are joined at the interface). It follows from this assumption that at the interface the stress normal to the interface and the displacement are continuous, i.e.,

$$(u_0 - u_1)_{r=a} = 0 \qquad (w_0 - w_1)_{r=a} = 0 \qquad (\sigma_{r0} - \sigma_{r1})_{r=a} = 0 \qquad (\tau_{rz0} - \tau_{rz1})_{r=a} = 0 \qquad (1.5)$$

Since the cylinder is assumed to be free and there are no surface forces, the normal and tangential forces at the surface must be zero, i.e.,

$$(\sigma_{r0})_{r=b} = 0 \qquad (\tau_{rz0})_{r=b} = 0 \qquad (\sigma_{zi})_{z=\pm H/2} = 0 \qquad (\tau_{rzi})_{z=\pm H/2} = 0 \qquad (i = 0, 1) \qquad (1.6)$$

The general solution of the system of equations (1.1) in the shell and in the core will be sought in the form

$$\left. \begin{array}{l} u_i = u_{i1} + u_{i2} \\ w_i = w_{i1} + w_{i2} \\ (i = 0, 1) \end{array} \right\} \qquad (1.7)$$

where u_{i1} and w_{i1} are the general solutions of (1.1) for $T = 0$, and u_{i2} and w_{i2} are arbitrary, particular solutions of (1.1).

The stresses are also written as sums:

$$\left. \begin{array}{l} \sigma_{ri} = \sigma_{ri1} + \sigma_{ri2} \\ \sigma_{\theta i} \equiv \sigma_{\theta i1} + \sigma_{\theta i2} \\ \sigma_{zi} = \sigma_{zi1} + \sigma_{zi2} \\ \tau_{rzi} = \tau_{rzi1} + \tau_{rzi2} \end{array} \right\} \qquad (1.8)$$

where σ_{ri1}, $\sigma_{\theta i1}$, σ_{zi1}, and τ_{rzi1} are determined by formulas (1.4) for $T = 0$, $u_i = u_{i1}$, and $w_i = w_{i1}$, while σ_{ri2}, $\sigma_{\theta i2}$, σ_{zi2}, and τ_{rzi2} are determined by the same formulas for $u_i = u_{i2}$ and $w_i = w_{i2}$. The general solutions u_{i1}, w_{i1} ($i = 0, 1$) in the shell and in the core are obtained by using the Love displacement functions $L = L_0(r, z)$ and $L = L_1(r, z)$, which satisfy the biharmonic equation

$$\Delta \Delta L(r, z) = 0$$

It is known that u_{i1} and w_{i1} are given in terms of the variables $L_i(r, z)$ by the formulas [7]

$$\left. \begin{array}{l} u_{i1} = -\dfrac{1 + \nu_i}{E_i} \dfrac{\partial^2 L_i}{\partial r \partial z} \\[12pt] w_{i1} = \dfrac{1 + \nu_i}{E} \left[2(1 - \nu_i)\Delta L_i - \dfrac{\partial^2 L_i}{\partial z^2} \right] \\[12pt] (i = 0, 1) \end{array} \right\} \qquad (1.9)$$

8

From (1.4), for $T = 0$, $u_i = u_{i1}$, and $w_i = w_{i1}$, we obtain the stresses in term of the $L_i(r, z)$:

$$
\left.
\begin{aligned}
\sigma_{ri1} &= \frac{\partial}{\partial z}\left(\nu_i\Delta L_i - \frac{\partial^2 L_i}{\partial r^2}\right) \\[2mm]
\sigma_{\theta i1} &= \frac{\partial}{\partial z}\left(\nu_i\Delta L_i - \frac{1}{r}\frac{\partial L_i}{\partial r}\right) \\[2mm]
\sigma_{zi1} &= \frac{\partial}{\partial z}\left[(2 - \nu_i)\Delta L_i - \frac{\partial^2 L_i}{\partial z^2}\right] \\[2mm]
\tau_{rzi1} &= \frac{\partial}{\partial r}\left[(1 - \nu_i)\Delta L_i - \frac{\partial^2 L_i}{\partial z^2}\right] \\[2mm]
&(i = 0, 1)
\end{aligned}
\right\}
\tag{1.10}
$$

The displacements $L = L_i(r, z)$ $(i = 0, 1)$ are expressed so that the general solutions of the system (1.1) and the general solutions for the stresses (1.8) satisfy the conditions (1.5) at the interface and the boundary conditions (1.6). Since u_{i2} and w_{i2} are any particular solutions of the system (1.1), we write them in the form

$$
\left.
\begin{aligned}
u_{i2} &= 0 \\[2mm]
w_{i2} &= \frac{1 + \nu_i}{1 - \nu_i}\int a_i(T)T(z)dz \\[2mm]
&(i = 0, 1)
\end{aligned}
\right\}
\tag{1.11}
$$

From (1.4), with $u_i = u_{i2}$ and $w_i = w_{i2}$, we obtain the stresses

$$
\left.
\begin{aligned}
\sigma_{ri2} = \sigma_{\theta i2} &= -\frac{E_i}{1 - \nu_i}a_iT \\[2mm]
\sigma_{zi2} = \tau_{rzi2} &= 0 \\[2mm]
&(i = 0, 1)
\end{aligned}
\right\}
\tag{1.12}
$$

The Thermally Stressed State of a Composite Cylinder with Temperature Function $\eta(z)$

In this chapter we assume that the thermal-expansion coefficients a_0 and a_1 are temperature-independent, i.e., that they are constant. Let the temperature field of a composite cylinder be given by (0.4). Then continue the function (0.4) to the interval $[-H, H]$ so that it is odd, and so that its Fourier expansion in this interval [6] is

$$\eta_1(z) = \frac{2}{\pi} \sum_{k=0}^{\infty} \frac{1}{2k + 1} \sin \frac{(2k + 1)\pi z}{H} \qquad (2.1)$$

We write the displacements in the shell and in the core in the form

$$L_0(r, z) = \sum_{k=0}^{\infty} [A_k I_0(n_k r) + B_k n_k r I_1(n_k r) + C_k K_0(n_k r) + D_k n_k r K_1(n_k r)] \cos n_k z$$

$$L_1(r, z) = \sum_{k=0}^{\infty} [E_k I_0(n_k r) + F_k n_k r I_1(n_k r)] \cos n_k z$$

where I_0, I_1, K_0, and K_1 are Bessel functions of an imaginary argument, A_k, B_k, C_k, D_k, E_k, and F_k are arbitrary constants, and $n_k = (2k + 1)\pi/H$.

The displacements and stresses in the shell obtained from (1.7) to (1.12) with the function (0.4) and the conditions $a_0 = \text{const}$ and $a_1 = \text{const}$ of this chapter are

$$\frac{u_0}{\eta_1} = \frac{1 + \nu_0}{E_0} \sum_{k=0}^{\infty} n_k^2 [A_k I_1(n_k r) + B_k n_k r I_0(n_k r) - C_k K_1(n_k r) - D_k n_k r K_0(n_k r)] \sin n_k z$$

$$\frac{w_0}{\eta_1} = \frac{1 + \nu_0}{E_0} \sum_{k=0}^{\infty} n_k^2 \{A_k I_0(n_k r) + B_k [4(1 - \nu_0)I_0(n_k r) + n_k r I_1(n_k r)] + C_k K_0(n_k r)$$

$$+ D_k [-4(1 - \nu_0)K_0(n_k r) + n_k r K_1(n_k r)] \} \cos n_k z + \frac{1 + \nu_0}{1 - \nu_0} a_0 \int \eta_1(z) dz$$

$$\frac{\sigma_{r0}}{\eta_1} = \sum_{k=0}^{\infty} n_k^3 \left\{ A_k \left[I_0(n_k r) - \frac{1}{n_k r} I_1(n_k r) \right] + B_k [(1 - 2\nu_0)I_0(n_k r) + n_k r I_1(n_k r)] \right.$$

10

$$+ C_k\left[K_0(n_k r) + \frac{1}{n_k r}K_1(n_k r)\right]$$

$$- D_k[(1 - 2\nu_0)K_0(n_k r) - n_k r K_1(n_k r)]\Bigg\}\sin n_k z - \frac{a_0 E_0}{1 - \nu_0}\eta_1(z) \qquad (2.2)$$

$$\frac{\sigma_{\theta 0}}{\eta_1} = \sum_{k=0}^{\infty} n_k^3\left[A_k\frac{1}{n_k r}I_1(n_k r) + B_k(1 - 2\nu_0)I_0(n_k r) - C_k\frac{1}{n_k r}K_1(n_k r)\right.$$

$$\left.- D_k(1 - 2\nu_0)K_0(n_k r)\right]\sin n_k z - \frac{a_0 E_0}{1 - \nu_0}\eta_1(z)$$

$$\frac{\sigma_{z 0}}{\eta_1} = -\sum_{k=0}^{\infty} n_k^3\{A_k I_0(n_k r) + B_k[2(2 - \nu_0)I_0(n_k r) + n_k r I_1(n_k r)] + C_k K_0(n_k r)$$

$$- D_k[2(2 - \nu_0)K_0(n_k r) - n_k r K_1(n_k r)]\}\sin n_k z$$

$$\frac{\tau_{rz 0}}{\eta_1} = \sum_{k=0}^{\infty} n_k^3\{A_k I_1(n_k r) + B_k[2(1 - \nu_0)I_1(n_k r) + n_k r I_0(n_k r)] - C_k K_1(n_k r)$$

$$+ D_k[2(1 - \nu_0)K_1(n_k r) - n_k r K_0(n_k r)]\}\cos n_k z$$

and those in the core are

$$\frac{u_1}{\eta_1} = \frac{1 + \nu_1}{E_1}\sum_{k=0}^{\infty} n_k^2[E_k I_1(n_k r) + F_k n_k r I_0(n_k r)]\sin n_k z$$

$$\frac{w_1}{\eta_1} = \frac{1 + \nu_1}{E_1}\sum_{k=0}^{\infty} n_k^2\{E_k I_0(n_k r) + F_k[4(1 - \nu_1)I_0(n_k r) + n_k r I_1(n_k r)]\}\cos n_k z$$

$$+ \frac{1 + \nu_1}{1 - \nu_1}a_1\int\eta_1(z)dz$$

$$\frac{\sigma_{r 1}}{\eta_1} = \sum_{k=0}^{\infty} n_k^3\left\{E_k\left[I_0(n_k r) - \frac{1}{n_k r}I_1(n_k r)\right] + F_k[(1 - 2\nu_1)I_0(n_k r) + n_k r I_1(n_k r)]\right\}\sin n_k z \qquad (2.3)$$

$$- \frac{a_1 E_1}{1 - \nu_1}\eta_1(z)$$

$$\frac{\sigma_{\theta 1}}{\eta_1} = \sum_{k=0}^{\infty} n_k^3\left[E_k\frac{1}{n_k r}I_1(n_k r) + F_k(1 - 2\nu_1)I_0(n_k r)\right]\sin n_k z - \frac{a_1 E_1}{1 - \nu_1}\eta_1(z)$$

$$\frac{\sigma_{z 1}}{\eta_1} = -\sum_{k=0}^{\infty} n_k^3\{E_k I_0(n_k r) + F_k[2(2 - \nu_1)I_0(n_k r) + n_k r I_1(n_k r)]\}\sin n_k z$$

11

$$\frac{\tau_{rz1}}{\eta_1} = \sum_{k=0}^{\infty} n_k^3 \{ E_k I_1(n_k r) + F_k[2(1 - \nu_1)I_1(n_k r) + n_k r I_0(n_k r)]\} \cos n_k z$$

For the general solutions (2.2) and (2.3) in the shell and core to satisfy conditions (1.5) at the interface and to satisfy the boundary conditions, the arbitrary constants A_k, B_k, C_k, D_k, E_k, and F_k must be obtained from the following set of six linear equations:

$$A_k I_1(n_k b) + B_k[2(1 - \nu_0)I_1(n_k b) + n_k b I_0(n_k b)] - C_k K_1(n_k b)$$
$$+ D_k[2(1 - \nu_0)K_1(n_k b) - n_k b K_0(n_k b)] = 0$$

$$A_k I_1(n_k a) + B_k[2(1 - \nu_0)I_1(n_k a) + n_k a I_0(n_k a)] - E_k I_1(n_k a)$$
$$- F_k[2(1 - \nu_1)I_1(n_k a) + n_k a I_0(n_k a)]$$
$$- C_k K_1(n_k a) + D_k[2(1 - \nu_0)K_1(n_k a) - n_k a K_0(n_k a)] = 0$$

$$A_k\left[I_0(n_k b) - \frac{1}{n_k b}I_1(n_k b)\right] + B_k[(1 - 2\nu_0)I_0(n_k b) + n_k b I_1(n_k b)]$$
$$+ C_k\left[K_0(n_k b) + \frac{1}{n_k b}K_1(n_k b)\right]$$
$$- D_k[(1 - 2\nu_0)K_0(n_k b) - n_k b K_1(n_k b)] = M_0$$

$$A_k\left[I_0(n_k a) - \frac{1}{n_k a}I_1(n_k a)\right] + B_k[(1 - 2\nu_0)I_0(n_k a) + n_k a I_1(n_k a)]$$
$$- E_k\left[I_0(n_k a) - \frac{1}{n_k a}I_1(n_k a)\right] - F_k[(1 - 2\nu_1)I_0(n_k r) + n_k a I_1(n_k a)] \tag{2.4}$$
$$+ C_k\left[K_0(n_k a) + \frac{1}{n_k a}K_1(n_k a)\right] - D_k[(1 - 2\nu_0)K_0(n_k a) - n_k a K_1(n_k a)] = M_0 - M_1$$

$$A_k \frac{1 + \nu_0}{E_0}I_0(n_k a) + B_k \frac{1 + \nu_0}{E_0}n_k a I_0(n_k a) - C_k \frac{1 + \nu_1}{E_1}I_1(n_k a) - F_k \frac{1 + \nu_1}{E_1}n_k a I_0(n_k a)$$
$$- C_k \frac{1 + \nu_0}{E_0}K_1(n_k a) - D_k \frac{1 + \nu_0}{E_0}n_k a K_0(n_k a) = 0$$

$$A_k \frac{1 + \nu_0}{E_0}I_0(n_k a) + B_k \frac{1 + \nu_0}{E_0}[4(1 - \nu_0)I_0(n_k a) + n_k a I_1(n_k a)] - E_k \frac{1 + \nu_1}{E_1}I_0(n_k a)$$
$$- F_k \frac{1 + \nu_1}{E_1}[4(1 - \nu_1)I_0(n_k a) + n_k a I_1(n_k a)] + C_k \frac{1 + \nu_0}{E_0}K_0(n_k a)$$
$$+ D_k \frac{1 + \nu_0}{E_0}[-4(1 - \nu_0)K_0(n_k a) + n_k a K_1(n_k a)] = \frac{1 + \nu_0}{E_0}M_0 - \frac{1 + \nu_1}{E_1}M_1$$

where

$$M_0 = \frac{a_0 E_0}{1 - \nu_0}\frac{2}{\pi(2k + 1)n_k^3} \qquad M_1 = \frac{a_1 E_1}{1 - \nu_1}\cdot\frac{2}{\pi(2k + 1)n_k^3}$$

To study the uniform convergence of the series (2.2) and (2.3) and to improve their rate of convergence, we must obtain the solution of the system (2.4) in the general (literal) form. Such a solution

12

is obtained by using Cramer's rule. After we have carried out some elementary transformations of the determinants, we obtain the solution of (2.4) in the form

$$A_k = \frac{M_0}{I_0(n_k b)}\frac{\Delta_1}{\Delta} \qquad B_k = \frac{M_0}{n_k b I_0(n_k b)}\frac{\Delta_2}{\Delta} \qquad E_k = -\frac{M_0}{I_0(n_k b)}\frac{\Delta_3}{\Delta}$$

$$F_k = \frac{M_0}{n_k b I_0(n_k b)}\frac{\Delta_4}{\Delta} \qquad C_k = \frac{M_0}{K_0(n_k a)}\cdot\frac{\Delta_5}{\Delta} \qquad D_k = \frac{M_0}{n_k b K_0(n_k a)}\frac{\Delta_6}{\Delta}$$

$$(2.5)$$

where Δ, Δ_1, Δ_2, Δ_3, Δ_4, Δ_5, and Δ_6 are the following determinants:

$$\Delta = \begin{vmatrix} \xi_2 & 1 & 0 & 0 & -\beta_3 & -\beta_1 \\ \xi_3 & \mu\xi_1 & \xi_3 & \mu\xi_1 & -\beta_2 & -\mu \\ 1 & s_0\xi_2 & 0 & 0 & \beta_1 & s_0\beta_3 \\ \xi_1 & q_0\xi_3 & \xi_1 & q_1\xi_3 & 1 & q_0\beta_2 \\ \xi_3 & \mu\xi_1 - \omega_0\xi_3 & \delta_0\xi_3 & \delta_0(\mu\xi_1 - \omega_1\xi_3) & -\beta_2 & -\mu - \omega_0\beta_2 \\ \xi_1 & \omega_0\xi_1 + \mu\xi_3 & \delta_0\xi_1 & \delta_0(\omega_1\xi_1 + \mu\xi_3) & 1 & -\omega_0 + \mu\beta_2 \end{vmatrix}$$

$$\Delta_1 = \begin{vmatrix} 0 & 1 + \omega_0\xi_2 & 0 & 0 & -\beta_3 & -\beta_1 \\ 0 & \omega_0\xi_3 + \mu\xi_1 & \xi_3 & \mu\xi_1 & -\beta_2 & -\mu \\ 1 & \omega_0 + s_0\xi_2 & 0 & 0 & \beta_1 & s_0\beta_3 \\ 1 - m & \omega_0\xi_1 + q_0\xi_3 & \xi_1 & q_1\xi_3 & 1 & q_0\beta_2 \\ 0 & \mu\xi_1 & \delta_0\xi_3 & \delta_0(\mu\xi_1 - \omega_1\xi_3) & -\beta_2 & -\mu - \omega_0\beta_2 \\ 1 - m\delta_0 & 2\omega_0\xi_1 + \mu\xi_3 & \delta_0\xi_1 & \delta_0(\omega_1\xi_1 + \mu\xi_3) & 1 & -\omega_0 + \mu\beta_2 \end{vmatrix}$$

$$\Delta_2 = \begin{vmatrix} \xi_2 & 0 & 0 & 0 & -\beta_3 & -\beta_1 \\ \xi_3 & 0 & \xi_3 & \mu\xi_1 & -\beta_2 & -\mu \\ 1 & 1 & 0 & 0 & \beta_1 & s_0\beta_3 \\ \xi_1 & 1 - m & \xi_1 & q_1\xi_3 & 1 & q_0\beta_2 \\ \xi_3 & 0 & \delta_0\xi_3 & \delta_0(\mu\xi_1 - \omega_1\xi_3) & -\beta_2 & -\mu - \omega_0\beta_2 \\ \xi_1 & 1 - m\delta_0 & \delta_0\xi_1 & \delta_0(\omega_1\xi_1 + \mu\xi_3) & 1 & -\omega_0 + \mu\beta_2 \end{vmatrix}$$

$$\Delta_3 = \begin{vmatrix} \xi_2 & 1 & 0 & 0 & -\beta_3 & -\beta_1 \\ \xi_3 & \mu\xi_2 & 0 & \omega_1\xi_3 + \mu\xi_1 & -\beta_2 & -\mu \\ 1 & s_0\xi_2 & 1 & 0 & \beta_1 & s_0\beta_3 \\ \xi_1 & q_0\xi_3 & 1 - m & \omega_1\xi_1 + q_1\xi_3 & 1 & q_0\beta_2 \\ \xi_3 & \mu\xi_1 - \omega_0\xi_3 & 0 & \delta_0\mu\xi_1 & -\beta_2 & -\mu - \omega_0\beta_2 \\ \xi_1 & \omega_0\xi_1 + \mu\xi_3 & 1 - m\delta_0 & \delta_0(2\omega_1\xi_1 + \mu\xi_3) & 1 & -\omega_0 + \mu\beta_2 \end{vmatrix}$$

$$\Delta_4 = \begin{vmatrix} \xi_2 & 1 & 0 & 0 & -\beta_3 & -\beta_1 \\ \xi_3 & \mu\xi_1 & \xi_3 & 0 & -\beta_2 & -\mu \\ 1 & s_0\xi_2 & 0 & 1 & \beta_2 & s_0\beta_3 \\ \xi_1 & q_0\xi_3 & \xi_1 & 1 - m & 1 & q_0\beta_2 \\ \xi_3 & \mu\xi_1 - \omega_0\xi_3 & \delta_0\xi_3 & 0 & -\beta_2 & -\mu - \omega_0\beta_2 \\ \xi_1 & \omega_0\xi_1 + \mu\xi_3 & \delta_0\xi_1 & 1 - m\delta_0 & 1 & -\omega_0 + \mu\beta_2 \end{vmatrix}$$

$$\Delta_5 = \begin{vmatrix} \xi_2 & 1 & 0 & 0 & 0 & \omega_0\beta_3 - \beta_1 \\ \xi_3 & \mu\xi_1 & \xi_3 & \mu\xi_1 & 0 & \omega_0\beta_2 - \mu \\ 1 & s_0\xi_2 & 0 & 0 & 1 & -\omega_0\beta_1 + s_0\beta_2 \\ \xi_1 & q_0\xi_3 & \xi_1 & q_1\xi_3 & 1 - m & -\omega_0 + q_0\beta_2 \\ \xi_3 & \mu\xi_1 - \omega_0\xi_3 & \delta_0\xi_3 & \delta_0(\mu\xi_1 - \omega_1\xi_3) & 0 & -\mu \\ \xi_1 & \omega_0\xi_1 + \mu\xi_3 & \delta_0\xi_1 & \delta_0(\omega_1\xi_1 + \mu\xi_3) & 1 - m\delta_0 & -2\omega_0 + \mu\beta_2 \end{vmatrix}$$

13

$$\Delta_6 = \begin{vmatrix} \xi_2 & 1 & 0 & 0 & -\beta_3 & 0 \\ \xi_3 & \mu\xi_1 & \xi_3 & \mu\xi_3 & -\beta_2 & 0 \\ 1 & s_0\xi_2 & 0 & 0 & \beta_2 & 1 \\ \xi_1 & q_0\xi_3 & \xi_1 & q_1\xi_3 & 1 & 1-m \\ \xi_3 & \mu\xi_1-\omega_0\xi_3 & \delta_0\xi_3 & \delta_0(\mu\xi_1-\omega_1\xi_3) & -\beta_2 & 0 \\ \xi_1 & \omega_0\xi_1+\mu\xi_3 & \delta_0\xi_1 & \delta_0(\omega_1\xi_1+\mu\xi_3) & 1 & 1-m\delta_0 \end{vmatrix}$$

where

$$\xi_1 = \frac{I_0(n_k a)}{I_0(n_k b)} \qquad \xi_2 = \frac{I_1(n_k b)}{I_0(n_k b)} \qquad \xi_3 = \frac{I_1(n_k a)}{I_0(n_k b)} \qquad \beta_1 = \frac{K_0(n_k b)}{K_0(n_k a)}$$

$$\beta_2 = \frac{K_1(n_k a)}{K_0(n_k a)} \qquad \beta_3 = \frac{K_1(n_k b)}{K_0(n_k a)} \qquad \mu = \frac{a}{b} \qquad s_0 = 1 + \frac{2(1-\nu_0)}{(n_k b)^2}$$

$$q_0 = \mu + \frac{2(1-\nu_0)}{\mu(n_k b)^2} \qquad q_1 = \mu + \frac{2(1-\nu_1)}{\mu(n_k b)^2} \qquad \omega_i = \frac{2(1-\nu_i)}{n_k b} \ (i = 0, 1) \qquad (2.6)$$

$$\delta_0 = \frac{G_0}{G_1} \qquad G_i = \frac{E_i}{2(1+\nu_i)} \ (i = 0, 1)$$

$$m = \frac{M_1}{M_0} = \frac{a_1 E_1}{1-\nu_0} : \frac{a_0 E_0}{1-\nu_0} = \frac{1-\nu_0}{1-\nu_1}\gamma\delta \qquad \gamma = \frac{a_1}{a_0} \qquad \delta = \frac{E_1}{E_0}$$

The determinants are expanded in their general form.

For the constants A_k, B_k, C_k, D_k, E_k, and F_k satisfying the system (2.4) and obtained from (2.5), all the equations in (1.5) and (1.6) are satisfied except

$$\left(\frac{\sigma_{zi}}{\eta_1}\right)_{z=\pm H/2} = 0 \ (i = 0, 1)$$

This last equality is not satisfied exactly, but the axial stresses $\frac{\sigma_{zi}}{\eta_1}$ at the ends compensate one another, i.e., their resultant is zero. This means that

$$\int_0^a \frac{\sigma_{z1}}{\eta_1} r\, dr + \int_a^b \frac{\sigma_{z0}}{\eta_1} r\, dr = 0 \qquad (2.7)$$

According to the Saint-Venant principle, the effect of stresses at the ends that are in equilibrium will not be felt at a certain distance from the ends. From the condition that the internal and external forces are in equilibrium, the principal force vector along any axis orthogonal to the cross section of the cylinder must be zero. Since there are no external forces and since the tangential stresses in orthogonal axes of a cross section are in equilibrium from the symmetry conditions, it follows that (2.7) must be satisfied for any z. We will show that (2.7) is, in fact, satisfied.

For $z = 0$ this is obvious, since $\frac{\sigma_{z1}}{\eta_1}(r, 0) = \frac{\sigma_{z0}}{\eta_1}(r, 0) = 0$ for all r except $r = a$ and $r = b$. We substitute the series (2.2) and (2.3) for $\frac{\sigma_{z0}}{\eta_1}$ and $\frac{\sigma_{z1}}{\eta_1}$ in (2.7), integrate term by term, and write the result as one series. This operation is legitimate since the series for $\frac{\sigma_{z0}}{\eta_1}$ and $\frac{\sigma_{z1}}{\eta_1}$ converge uniformly and absolutely for $|z| > 0$.

14

Using the relations

$$\int r I_0(n_k r) dr = \frac{r}{n_k} I_1(n_k r)$$

$$\int n_k r^2 I_1(n_k r) dr = \frac{r}{n_k}[n_k r I_0(n_k r) - 2I_1(n_k r)]$$

$$\int r K_0(n_k r) dr = -\frac{r}{n_k} K_1(n_k r)$$

$$\int n_k r^2 K_1(n_k r) dr = -\frac{r}{n_k}[n_k r K_0(n_k r) + 2K_1(n_k r)]$$

we obtain

$$\int_0^a \frac{\sigma_{z1}}{\eta_1} r\, dr + \int_a^b \frac{\sigma_{z0}}{\eta_1} r\, dr = -\sum_{k=0}^{\infty} n_k^2 \Big\{ b\{A_k I_1(n_k b) + B_k[2(1 - \nu_0)I_1(n_k b) + n_k b I_0(n_k b)] - C_k K_1(n_k b)$$

$$+ D_k[2(1 - \nu_0)K_1(n_k b) - n_k b K_0(n_k b)]\} - a\{A_k I_1(n_k a) + B_k[2(1 - \nu_0)I_1(n_k a) + n_k a I_0(n_k a)]$$

$$- E_k I_1(n_k a) - F_k[2(1 - \nu_1)I_1(n_k a) + n_k a I_0(n_k a)] - C_k K_1(n_k a)$$

$$+ D_k[2(1 - \nu_0)K_1(n_k a) - n_k a K_0(n_k a)]\}\Big\} \sin n_k z = 0$$

since, according to the first and second equations of (2.4), the expressions in the small braces are zero.

For the normal stresses $\frac{\sigma_{r0}}{\eta_1}$, $\frac{\sigma_{\theta 0}}{\eta_1}$, and $\frac{\sigma_{z0}}{\eta_1}$ in the shell, the series (2.2) converge nonuniformly in $0 \le z \le H/2$, $a \le r \le b$ and define a function of r and z that is discontinuous for $r = a$, $z = 0$ and $r = b$, $z = 0$.

For the normal stresses $\frac{\sigma_{r1}}{\eta_1}$, $\frac{\sigma_{\theta 1}}{\eta_1}$, and $\frac{\sigma_{z1}}{\eta_1}$ in the core, the series (2.3) converge nonuniformly in $0 \le z \le H/2$, $0 \le r \le a$ and define a function of r and z that is discontinuous for $r = z$, $z = 0$.

By separating out the known series (2.1), we can replace the series (2.2) and (2.3) by series that converge uniformly in the region under consideration, and we can simultaneously increase the rate of convergence.

We will investigate the uniform convergence of the series (2.2) and (2.3) for the stresses in the shell and core, and we will accelerate their convergence.

The Convergence of (2.2)

We introduce the notation

$$\Delta_{r0k} = \frac{1}{M_0}\Big\{A_k\Big[I_0(n_k r) - \frac{1}{n_k r} I_1(n_k r)\Big] + B_k[(1 - 2\nu_0)I_0(n_k r) + n_k r I_1(n_k r)] + C_k\Big[K_0(n_k r) + \frac{1}{n_k r} K_1(n_k r)\Big]$$

$$- D_k[(1 - 2\nu_0)K_0(n_k r) - n_k r K_1(n_k r)]\Big\}$$

$$\Delta_{\theta 0k} = \frac{1}{M_0}\Big[A_k \frac{1}{n_k r} I_1(n_k r) + B_k(1 - 2\nu_0)I_0(n_k r) - C_k \frac{1}{n_k r} K_1(n_k r) - D_k(1 - 2\nu_0)K_0(n_k r)\Big]$$

15

$$\Delta_{z0k} = \frac{1}{M_0}\{A_k I_0(n_k r) + B_k[2(2 - \nu_0)I_0(n_k r) + n_k r I_1(n_k r)] + C_k K_0(n_k r)$$

$$- D_k[2(2 - \nu_0)K_0(n_k r) - n_k r K_1(n_k r)]\}$$

$$\Delta_{rz0k} = \frac{1}{M_0}\{A_k I_1(n_k r) + B_k[2(1 - \nu_0)I_1(n_k r) + n_k r I_0(n_k r)] - C_k K_1(n_k r)$$

$$+ D_k[2(1 - \nu_0)K_1(n_k r) - n_k r K_0(n_k r)]\}$$

Then the stresses given by (2.2) can be expressed as follows:

$$\frac{\sigma_{r0}}{\eta_1} = \frac{2a_0 E_0}{\pi(1 - \nu_0)} \sum_{k=0}^{\infty} \Delta_{r0k} \cdot \frac{\sin n_k z}{2k + 1} - \frac{a_0 E_0}{1 - \nu_0} \eta_1(z) \qquad (2.8)$$

$$\frac{\sigma_{\theta 0}}{\eta_1} = \frac{2a_0 E_0}{\pi(1 - \nu_0)} \sum_{k=0}^{\infty} \Delta_{\theta.0k} \cdot \frac{\sin n_k z}{2k + 1} - \frac{a_0 E_0}{1 - \nu_0} \eta_1(z) \qquad (2.9)$$

$$\frac{\sigma_{z0}}{\eta_1} = \frac{2a_0 E_0}{\pi(1 - \nu_0)} \sum_{k=0}^{\infty} \Delta_{z0k} \cdot \frac{\sin n_k z}{2k + 1} \qquad (2.10)$$

$$\frac{\tau_{rz0k}}{\eta_1} = \frac{2a_0 E_0}{\pi(1 - \nu_0)} \sum_{k=0}^{\infty} \Delta_{rz0k} \cdot \frac{\cos n_k z}{2k + 1} \qquad (2.11)$$

We will obtain similar expressions for Δ_{r0k}, $\Delta_{\theta 0k}$, Δ_{z0k}, and Δ_{rz0k} for $a \leq r \leq b$ and $k \to \infty$.

We consider three cases.

1. $r = a$

$$\Delta_{r0k} \sim \frac{1}{M + 1} + \theta + \frac{C_0}{2k + 1} \qquad C_0 = \text{const}$$

$$\Delta_{\theta 0k} \sim \frac{2\nu_0}{M + 1} + \frac{C_1}{2k + 1} \qquad C_1 = \text{const}$$

$$\Delta_{z0k} \sim -\frac{1}{M + 1} + \theta + \frac{C_2}{2k + 1} \qquad C_2 = \text{const}$$

$$\Delta_{rz0k} \sim -\theta + \frac{C_3}{2k + 1} \qquad C_3 = \text{const}$$

where

$$\theta = \frac{1}{2}\left[\frac{M}{M + 1} - \frac{m(N - 1)}{N}\right] \qquad M = \frac{4G_1(1 - \nu_0)}{G_0 - G_1}$$

$$N = 1 - \frac{4G_0(1 - \nu_1)}{G_0 - G_1} \qquad G_i = \frac{E_i}{2(1 + \nu_i)} \quad (i = 0, 1) \qquad (2.12)$$

In this case the series (2.8), (2.9), and (2.10) converge rather slowly and nonuniformly in $[0, H/2]$. Ex-

16

tracting the known series (2.1) from these series, we obtain

$$\frac{\sigma_{r0}}{\eta_1} = \frac{2a_0E_0}{\pi(1-\nu_0)}\sum_{k=0}^{\infty}\left(\Delta_{r0k} - \frac{1}{M+1} - \theta\right)\frac{\sin n_k z}{2k+1} - \frac{a_0E_0}{1-\nu_0}\left(1 - \frac{1}{M+1} - \theta\right)\eta_1(z) \qquad (2.13)$$

$$\frac{\sigma_{\theta 0}}{\eta_1} = \frac{2a_0E_0}{\pi(1-\nu_0)}\sum_{k=0}^{\infty}\left(\Delta_{\theta 0k} - \frac{2\nu_0}{M+1}\right)\frac{\sin n_k z}{2k+1} - \frac{a_0E_0}{1-\nu_0}\left(1 - \frac{2\nu_0}{M+1}\right)\eta_1(z) \qquad (2.14)$$

$$\frac{\sigma_{z0}}{\eta_1} = -\frac{2a_0E_0}{\pi(1-\nu_0)}\sum_{k=0}^{\infty}\left(\Delta_{z0k} + \frac{1}{M+1} - \theta\right)\frac{\sin n_k z}{2k+1} + \frac{a_0E_0}{1-\nu_0}\left(\frac{1}{M+1} - \theta\right)\eta_1(z) \qquad (2.15)$$

It is easily seen that the series (2.13), (2.14), and (2.15), starting from some value of k, are majorized by the convergent series with the general term $C/(2k+1)^2$, $C = $ const, and so, from the Weierstrass criterion, they converge uniformly in $[0, H/2]$.

For $z = +0$, the normal stresses in the shell at the interface are

$$\frac{\sigma_{r0}}{\eta_1} = -\frac{a_0E_0}{2(1-\nu_0)}\cdot\left(1 - \frac{1}{M+1} - \theta\right) \qquad \frac{\sigma_{\theta 0}}{\eta_1} = -\frac{a_0E_0}{2(1-\nu_0)}\cdot\left(1 - \frac{2\nu_0}{M+1}\right)$$

$$\frac{\sigma_{z0}}{\eta_1} = \frac{a_0E_0}{2(1-\nu_0)}\cdot\left(\frac{1}{M+1} - \theta\right) \qquad (2.16)$$

We extract from the series (2.11) the known series

$$\sum_{k=0}^{\infty}\frac{1}{2k+1}\cos\frac{(2k+1)\pi z}{H} = -\tfrac{1}{2}\ln\tan\frac{\pi z}{2H} \qquad 0 < z < H \qquad (2.17)$$

and rewrite the tangential stress $\frac{\tau_{rz0}}{\eta_1}$ on the interface ($r = a$) in the form

$$\frac{\tau_{rz0}}{\eta_1} = \frac{2a_0E_0}{\pi(1-\nu_0)}\sum_{k=0}^{\infty}(\Delta_{rz0k} + \theta)\frac{\cos n_k z}{2k+1} + \frac{a_0E_0\theta}{\pi(1-\nu_0)}\ln\left|\tan\frac{\pi z}{2H}\right| \qquad \left(-\frac{H}{2} \le z \le \frac{H}{2}\right) \, (2.18)$$

From the Weierstrass criterion, the series here converges uniformly in the interval $[-H/2, H/2]$ since, starting from some value of k, it is majorized by the series with the general term $C/(2k+1)^2$, $C = $ const.

It follows from (2.18) that, for $r = a$, we have

$$\lim_{z\to\pm 0}\frac{\tau_{rz0}}{\eta_1} = \begin{cases} +\infty & \text{if } \theta < 0 \\ \tau = \text{const} & \text{if } \theta = 0 \\ -\infty & \text{if } \theta > 0 \end{cases}$$

2. $a < r < b$

$$\Delta_{r0k} \text{ and } \Delta_{z0k} \sim \sqrt{\frac{b}{r}}\,(b-r)n_k e^{(r-b)n_k} + \sqrt{\frac{a}{r}}\,\frac{n_k(r-a)}{M+1}e^{(a-r)n_k}$$

$$\Delta_{\theta 0k} \sim \left(\frac{b}{r} - 1 + 2\nu_0\right)\sqrt{\frac{b}{r}}\,e^{(r-b)n_k} + \left(\frac{a}{r} - 1 + 2\nu_0\right)\sqrt{\frac{a}{r}}\,\frac{1}{M+1}e^{(a-r)n_k}$$

17

$$\Delta_{rz0k} \sim \sqrt{\frac{b}{r}}(b-r)n_k e^{(r-b)n_k} - \sqrt{\frac{a}{r}}\frac{(r-a)n_k}{M+1}e^{(a-r)n_k}$$

Here the series (2.8), (2.9), (2.10), and (2.11) converge uniformly in z in the interval $[0, H/2]$ from the Weierstrass criterion, since

$$\left|\Delta_{i0k} \cdot \frac{\sin n_k z}{2k+1}\right| \leq \left|\frac{\Delta_{i0k}}{2k+1}\right| \qquad (i = r, \theta, z)$$

$$\left|\Delta_{rz0k}\frac{\cos n_k z}{2k+1}\right| \leq \left|\frac{\Delta_{rz0k}}{2k+1}\right|$$

and the series with general terms $|\Delta_{i0k}/(2k+1)|$ $(i = r, \theta, z, rz)$ converge. This follows from the fact that, according to the relation obtained above, the series with general terms equivalent to $|\Delta_{i0k}/(2k+1)|$ converge. For $z = +0$ and $a < r < b$, the normal stresses in the shell are

$$\frac{\sigma_{r0}}{\eta_1} = -\frac{a_0 E_0}{2(1-\nu_0)} \qquad \frac{\sigma_{\theta 0}}{\eta_1} = -\frac{a_0 E_0}{2(1-\nu_0)} \qquad \frac{\sigma_{z0}}{\eta_1} = 0 \qquad (2.19)$$

A comparison of (2.16) and (2.19) shows that the normal stresses in the shell are discontinuous for $z = +0$ and $r = a$ (on the interface between the shell and the core).

3. $r = b$

$$\Delta_{r0k} = 1$$

$$\Delta_{\theta 0k} \sim 2\nu_0 + \frac{C_4}{2k+1} \qquad C_4 = \text{const}$$

$$\Delta_{z0k} \sim -1 + \frac{C_5}{2k+1} \qquad C_5 = \text{const}$$

$$\Delta_{rz0k} = 0$$

On the surface of the cylinder $\frac{\sigma_{r0}}{\eta_1} = \frac{\tau_{rz0}}{\eta_1} = 0$, and the series (2.9) and (2.10) converge rather slowly and nonuniformly in $[0, H/2]$. Extracting the known series (2.1) from these series, we obtain

$$\frac{\sigma_{\theta 0}}{\eta_1} = \frac{2a_0 E_0}{\pi(1-\nu_0)}\sum_{k=0}^{\infty}(\Delta_{\theta 0k}-2\nu_0)\frac{\sin n_k z}{2k+1} - \frac{1-2\nu_0}{1-\nu_0}a_0 E_0\eta_1(z) \qquad (2.20)$$

$$\frac{\sigma_{z0}}{\eta_1} = -\frac{2a_0 E_0}{\pi(1-\nu_0)}\sum_{k=0}^{\infty}(\Delta_{z0k}+1)\frac{\sin n_k z}{2k+1} + \frac{a_0 E_0}{1-\nu_0}\eta_1(z) \qquad (2.21)$$

It is easily seen from the Weierstrass criterion that the series (2.20) and (2.21) converge uniformly in $[0, H/2]$, since they are majorized by the convergent series with the general term $C/(2k+1)^2$, $C = \text{const}$. For $z = +0$ and $r = b$, the normal stresses are

$$\frac{\sigma_{\theta 0}}{\eta_1} = -\frac{1-2\nu_0}{2(1-\nu_0)}a_0 E_0 \qquad \frac{\sigma_{z0}}{\eta_1} = \frac{a_0 E_0}{2(1-\nu_0)} \qquad \frac{\sigma_{r0}}{\eta_1} = 0 \qquad (2.22)$$

18

A comparison of (2.19) and (2.22) shows that the normal stresses in the shell are discontinuous for $z = +0$ and $r = b$, i.e., on the surface.

The Convergence of (2.3)

We write

$$\Delta_{r1k} = \frac{1}{M_0}\left\{E_k\left[I_0(n_k r) - \frac{1}{n_k r}I_1(n_k r)\right] + F_k[(1 - 2\nu_1)I_0(n_k r) + n_k r I_1(n_k r)]\right\}$$

$$\Delta_{\theta 1k} = \frac{1}{M_0}\left[E_k \frac{1}{n_k r}I_1(n_k r) + F_k(1 - 2\nu_1)I_0(n_k r)\right]$$

$$\Delta_{z1k} = \frac{1}{M_0}\{E_k I_0(n_k r) + F_k[2(2 - \nu_1)I_0(n_k r) + n_k r I_1(n_k r)]\}$$

$$\Delta_{rz1k} = \frac{1}{M_0}\{E_k I_1(n_k r) + F_k[2(1 - \nu_1)I_1(n_k r) + n_k r I_0(n_k r)]\}$$

Then the series (2.3) for the stresses in the core become

$$\frac{\sigma_{r1}}{\eta_1} = \frac{2a_0 E_0}{\pi(1 - \nu_0)}\sum_{k=0}^{\infty}\Delta_{r1k}\frac{\sin n_k z}{2k + 1} - m \cdot \frac{a_0 E_0}{1 - \nu_0}\eta_1(z) \qquad (2.23)$$

$$\frac{\sigma_{\theta 1}}{\eta_1} = \frac{2a_0 E_0}{\pi(1 - \nu_0)}\sum_{k=0}^{\infty}\Delta_{\theta 1k}\cdot\frac{\sin n_k z}{2k + 1} - m \cdot \frac{a_0 E_0}{1 - \nu_0}\eta_1(z) \qquad (2.24)$$

$$\frac{\sigma_{z1}}{\eta_1} = -\frac{2a_0 E_0}{\pi(1 - \nu_0)}\sum_{k=0}^{\infty}\Delta_{z1k}\cdot\frac{\sin n_k z}{2k + 1} \qquad (2.25)$$

$$\frac{\tau_{rz1}}{\eta_1} = \frac{2a_0 E_0}{\pi(1 - \nu_0)}\sum_{k=0}^{\infty}\Delta_{rz1k}\cdot\frac{\cos n_k z}{2k + 1} \qquad (2.26)$$

where $m = M_1/M_0 = a_1 E_1/(1 - \nu_1) : a_0 E_0/(1 - \nu_0)$ [see the notation (2.6)].

We now find the equivalent relations for Δ_{r1k}, $\Delta_{\theta 1k}$, Δ_{z1k}, and Δ_{rz1k} for $0 \le r \le a$ and $k \to \infty$. Again we consider three cases.

1. $r = 0$

$$\Delta_{r1k} \text{ and } \Delta_{\theta 1k} \sim \frac{m\sqrt{2\pi}}{2N}(n_k a)^{3/2}e^{-n_k a}$$

$$\Delta_{z1k} \sim \frac{-m\sqrt{2\pi}}{N}(n_k a)^{3/2}e^{-n_k a}$$

$$\Delta_{rz1k} = 0 \qquad \text{for all } k$$

We note that for $r = 0$ we have $\frac{\sigma_{r1}}{\eta_1} = \frac{\sigma_{\theta 1}}{\eta_1}, \frac{\tau_{rz1}}{\eta_1} = 0$.

19

Here the series (2.23) to (2.26) converge uniformly in z in the interval $[0, H/2]$ from the Weierstrass criterion [the proof is the same as in case 2 in the investigation of the series (2.8) to (2.11)]. For $z = +0$ and $r = 0$, the normal stresses are

$$\frac{\sigma_{r1}}{\eta_1} = \frac{\sigma_{\theta 1}}{\eta_1} = -\frac{m}{2(1 - \nu_0)} a_0 E_0 \qquad \frac{\sigma_{z1}}{\eta_1} = 0$$

2. $0 < r < a$

$$\Delta_{r1k}, \Delta_{z1k}, \Delta_{rz1k} \sim \frac{m}{N} \sqrt{\frac{a}{r}} n_k (a - r) e^{-(a-r)n_k}$$

$$\Delta_{\theta 1k} \sim \frac{m}{N} \left(\frac{a}{r} - 1 + 2\nu_1 \right) \sqrt{\frac{a}{r}} e^{-(a-r)n_k}$$

Here the series (2.23) to (2.26) converge uniformly in z in the interval $[0, H/2]$ from the Weierstrass criterion (again, the proof is the same as in case 2 above). For $z = +0$ and $0 < r < a$, the normal stresses in the core are

$$\frac{\sigma_{r1}}{\eta_1} = \frac{\sigma_{\theta 1}}{\eta_1} = -\frac{m}{2(1 - \nu_0)} a_0 E_0 \qquad \frac{\sigma_{z1}}{\eta_1} = 0 \qquad (2.27)$$

3. $r = a$

$$\Delta_{r1k} \sim \frac{m}{N} - \theta + \frac{C_1}{2k + 1} \qquad C_1 = \text{const}$$

$$\Delta_{\theta 1k} \sim \frac{2m\nu_1}{N} + \frac{C_2}{2k + 1} \qquad C_2 = \text{const}$$

$$\Delta_{z1k} \sim -\frac{m}{N} - \theta + \frac{C_3}{2k + 1} \qquad C_3 = \text{const}$$

$$\Delta_{rz1k} \sim -\theta + \frac{C_4}{2k + 1} \qquad C_4 = \text{const}$$

In this case the series (2.23) to (2.25) converge relatively slowly and nonuniformly in $[0, H/2]$. Extracting the known series (2.1), we obtain

$$\frac{\sigma_{r1}}{\eta_1} = \frac{2a_0 E_0}{\pi(1 - \nu_0)} \sum_{k=0}^{\infty} \left(\Delta_{r1k} - \frac{m}{N} + \theta \right) \frac{\sin n_k z}{2k + 1} - \frac{a_0 E_0}{1 - \nu_0} \left(m - \frac{m}{N} + \theta \right) \eta_1(z) \qquad (2.28)$$

$$\frac{\sigma_{\theta 1}}{\eta_1} = \frac{2a_0 E_0}{\pi(1 - \nu_0)} \sum_{k=0}^{\infty} \left(\Delta_{\theta 1k} - \frac{2m\nu_1}{N} \right) \frac{\sin n_k z}{2k + 1} - m \frac{a_0 E_0}{1 - \nu_0} \left(1 - \frac{2\nu_1}{N} \right) \eta_1(z) \qquad (2.29)$$

$$\frac{\sigma_{z1}}{\eta_1} = -\frac{2a_0 E_0}{\pi(1 - \nu_0)} \sum_{k=0}^{\infty} \left(\Delta_{z1k} + \frac{m}{N} + \theta \right) \frac{\sin n_k z}{2k + 1} + \frac{a_0 E_0}{1 - \nu_0} \left(\frac{m}{N} + \theta \right) \eta_1(z) \qquad (2.30)$$

For $r = a$, the series (2.28) to (2.30) converge uniformly in $[0, H/2]$ [the proof here is the same as that in case 1 in the investigation of the convergence of series (2.8) to (2.11)].

For $z = +0$, the normal stresses in the core at the interface ($r = a$) are

$$\frac{\sigma_{r1}}{\eta_1} = -\frac{a_0 E_0}{2(1 - \nu_0)}\left(m - \frac{m}{N} + \theta\right)$$

$$\frac{\sigma_{\theta 1}}{\eta_1} = -\frac{m}{2(1 - \nu_0)}\left(1 - \frac{2\nu_1}{N}\right)a_0 E_0 \qquad (2.31)$$

$$\frac{\sigma_{z1}}{\eta_1} = \frac{a_0 E_0}{2(1 - \nu_0)}\left(\frac{m}{N} + \theta\right)$$

A comparison of (2.27) with (2.31) shows that the normal stresses in the core are discontinuous for $z = +0$ and $r = a$.

We extract from the series (2.26) the known series (2.17) and rewrite the tangential stress $\frac{\tau_{rz1}}{\eta_1}$ on the interface ($r = a$) in the form

$$\frac{\tau_{rz1}}{\eta_1} = \frac{2a_0 E_0}{\pi(1 - \nu_0)} \sum_{k=0}^{\infty} (\Delta_{rz1k} + \theta)\frac{\cos n_k z}{2k + 1} + \frac{a_0 E_0 \theta}{\pi(1 - \nu_0)} \ln\left|\tan\frac{\pi z}{2H}\right| \qquad \left(-\frac{H}{2} \le z \le \frac{H}{2}\right) \qquad (2.32)$$

The series in this equation converges uniformly in $[-H/2, H/2]$ since, starting from some value of k, it is majorized by the convergent series with general term $C/(2k + 1)^2$, $C = $ const.

It follows from (2.32) that, for $r = a$, we have

$$\lim_{z \to \pm 0} \frac{\tau_{rz1}}{\eta_1} = \begin{cases} +\infty & \text{if } \theta < 0 \\ \tau = \text{const} & \text{if } \theta = 0 \\ -\infty & \text{if } \theta > 0 \end{cases}$$

In concluding the investigation of the convergence of the series (2.2) and (2.3), we note that the number of terms in the series that must be used to obtain a given accuracy in the stresses depends on the values of $|a - r|$ and $|b - r|$. When these quantities are small, the number of terms required is large. The series converge most rapidly for $r = 0$ and most slowly for $r = a$. To obtain the stresses with an error of less than 0.0005 in the first case it is sufficient to sum 30 terms. In the second case about 2000 terms are needed for values of $\mu = a/b$ from 0.7 up to 0.95.

We will express the stresses $\frac{\sigma_{ij}}{\eta_1}, \frac{\tau_{rzj}}{\eta_1}$ ($i = r, \theta, z; j = 0, 1$) as products of the stresses $\frac{\overline{\sigma}_{ij}}{\eta_1}, \frac{\overline{\tau}_{rzj}}{\eta_1}$ and $a_0 E_0$:

$$\frac{\sigma_{ij}}{\eta_1} = a_0 E_0 \frac{\overline{\sigma}_{ij}}{\eta_1}$$

$$\frac{\tau_{rzj}}{\eta_1} = a_0 E_0 \frac{\overline{\tau}_{rzj}}{\eta_1}$$

$$(i = r, \theta, z; j = 0, 1)$$

The stresses $\frac{\overline{\sigma}_{ij}}{\eta_1}, \frac{\overline{\tau}_{rzj}}{\eta_1}$ depend on the parameters $\mu = a/b$, $\delta = E_1/E_0$, $\gamma = a_1/a_0$, ν_1, ν_0, and the two independent variables r and z. The Poisson ratio for concrete was determined approximately, and in practice it can be assumed to be the same for all concretes and equal to $\nu = 1/6$. For calculations of stresses in composite concrete cylinders we therefore usually assume that $\nu_0 = \nu_1 = 1/6$. Hence,

$\underset{\eta_1}{\bar\sigma_{ij}}$ and $\underset{\eta_1}{\bar\tau_{rzj}}$ will depend on the three parameters μ, δ, γ, and the two independent variables r and z:

$$\underset{\eta_1}{\bar\sigma_{ij}} = \underset{\eta_1}{\bar\sigma_{ij}}(\mu, \gamma, \delta, r, z)$$

$$\underset{\eta_1}{\bar\tau_{rzj}} = \underset{\eta_1}{\bar\tau_{rzj}}(\mu, \gamma, \delta, r, z)$$

$$(i = r, \theta, z; \; j = 0, 1)$$

It follows from a theorem in similarity theory that the solutions of (2.2) and (2.3) will remain unchanged if r and z are replaced by r/b and z/b; thus

$$\left.\begin{aligned}
\underset{\eta_1}{\bar\sigma_{ij}}(\mu, \gamma, \delta, r, z) &= \underset{\eta_1}{\bar\sigma_{ij}}\left(\mu, \gamma, \delta, \frac{r}{b}, \frac{z}{b}\right) \\[2ex]
\underset{\eta_1}{\bar\tau_{rzj}}(\mu, \gamma, \delta, r, z) &= \underset{\eta_1}{\bar\tau_{rzj}}\left(\mu, \gamma, \delta, \frac{r}{b}, \frac{z}{b}\right)
\end{aligned}\right\} \tag{2.33}$$

We will not prove this; the proof is the same as for a homogeneous cylinder [6]. Calculations have also verified the truth of (2.33).

The stresses $\underset{\eta_1}{\bar\sigma_{ij}}(\mu, \gamma, \delta, r, z)$ and $\underset{\eta_1}{\bar\tau_{rzj}}(\mu, \gamma, \delta, r, z)$ $(i = r, \theta, z; \; j = 0, 1)$ were calculated on a BESM-2 computer. Not only the function $\eta_1(z)$ was calculated, but also the function $\eta_1(z - z_0)$. The following results were obtained for

$$|z_0| \leq \frac{H}{2} - p_0 \pi b \qquad H > 2 p_0 \pi b \tag{2.34}$$

with p_0 a constant depending on the accuracy of the calculation of the stresses; this constant is 0.8 when the absolute error is $|\Delta|_{\eta_1} = 0.0005$.

1. The stresses for the temperature function $\eta_1(z - z_0)$ are equal to the stresses for the function $\eta_1(z)$ translated by the same amount z_0 as the temperature:

$$\left.\begin{aligned}
\underset{\eta_1(z - z_0)}{\bar\sigma_{ij}}(\mu, \gamma, \delta, r, z) &= \underset{\eta_1(z)}{\bar\sigma_{ij}}(\mu, \gamma, \delta, r, z - z_0) \\[2ex]
\underset{\eta_1(z - z_0)}{\bar\tau_{rzj}}(\mu, \gamma, \delta, r, z) &= \underset{\eta_1(z)}{\bar\tau_{rzj}}(\mu, \gamma, \delta, r, z - z_0) \\[2ex]
(i = r, \theta, z; \; j = 0, 1)
\end{aligned}\right\} \tag{2.35}$$

2. For $|z - z_0| \geq p_0 \pi b$, the stresses (2.2) and (2.3) with an error less than $|\Delta|_{\eta_1}$ are independent of z and are:

for $z - z_0 \geq p_0 \pi b$:

$$\left.\begin{aligned}
\underset{\eta_1(z - z_0)}{\bar\sigma_{ij}}(\mu, \gamma, \delta, r, z) &= 0.5\underset{1}{\bar\sigma_{ij}}(\mu, \gamma, \delta, r) \\[2ex]
\underset{\eta_1(z - z_0)}{\bar\tau_{rzj}}(\mu, \gamma, \delta, r, z) &= 0.5\underset{1}{\bar\tau_{rzj}}(\mu, \gamma, \delta, r) \\[2ex]
(i = r, \theta, z; \; j = 0, 1)
\end{aligned}\right\} \tag{2.36}$$

22

for $z - \dot z_0 \leq -p_0\pi b$:

$$
\left.
\begin{aligned}
\underset{\substack{1 \\ \eta_1(z-z_0)}}{\overline{\sigma}_{ij}}(\mu, \gamma, \delta, r, z) &= -0.5\underset{1}{\overline{\sigma}}_{ij}(\mu, \gamma, \delta, r) \\[2ex]
\underset{\substack{1 \\ \eta_1(z-z_0)}}{\overline{\tau}_{rzj}}(\mu, \gamma, \delta, r, z) &= 0.5 \cdot \underset{1}{\overline{\tau}}_{rzj}(\mu, \gamma, \delta, r) \\[2ex]
(i = r, \theta, z; \; j &= 0, 1)
\end{aligned}
\right\}
\tag{2.37}
$$

where $\underset{1}{\overline{\sigma}}_{ij}(\mu, \gamma, \delta, r)$ and $\underset{1}{\overline{\tau}}_{rzj}(\mu, \gamma, \delta, r)$ $(i = r, \theta, z; \; j = 0, 1)$ are the thermal stresses in a composite

cylinder expressed as multiples of $a_0 E_0$ for the temperature function $T = 1$; this is the particular case of a temperature function $T = T(r)$. The stresses $\underset{1}{\sigma}_{ij}$ and $\underset{1}{\tau}_{rzj}$ for $T = 1$ are given in terms of $\underset{1}{\overline{\sigma}}_{ij}$ and

$\underset{1}{\overline{\tau}}_{rzj}$ by the formulas

$$
\underset{1}{\sigma}_{ij} = a_0 E_0 \underset{1}{\overline{\sigma}}_{ij}
$$

$$
\underset{1}{\tau}_{rzj} = a_0 E_0 \underset{1}{\overline{\tau}}_{rzj}
$$

$$
(i = r, \theta, z; \; j = 0, 1)
$$

The problem of the thermally stressed state of a cylinder with a temperature function $T = T(r)$ and with no surface forces was solved by Gatewood [8]. This problem is considered for a cylinder long enough so that it can be assumed to be plane at distances sufficiently far from its ends. In this case (plane deformation), however, the total axial deformation in a composite cylinder is constant $(\overline{\epsilon}_z = \overline{\epsilon}_{z0} = \overline{\epsilon}_{z1} = \text{const})$ and the axial stresses σ_{zj} $(j = 0, 1)$ are not zero.

Hence, with an error of less than 0.0005, our problem becomes a plane problem at the distance $0.8\pi b \approx 2.5b$ from the point z_0 of discontinuity of the function $\eta_1(z - z_0)$ where there is the maximum temperature gradient $\eta_1'(z - z_0)|_{z=z_0} = \delta(z - z_0)|_{z=z_0} = \delta(0) = \infty$ [$\delta(z)$ is the delta function]. In other words, orthogonal axes of a cross section remain orthogonal after deformation.

The solution of the plane problem for a composite cylinder with a temperature function $T = T(r)$ is obtained by the method of N. N. Muskhelishvili, which uses analytic functions of a complex variable, and for the particular case when $T = T_0 = \text{const}$, the solution is [8]

in the shell:

$$
\left.
\begin{aligned}
\frac{\sigma_{r0}}{T_0} &= A\left(1 - \frac{b^2}{r^2}\right) \\[2ex]
\frac{\sigma_{\theta 0}}{T_0} &= A\left(1 + \frac{b^2}{r^2}\right) \\[2ex]
\frac{\tau_{r\theta 0}}{T_0} &= \frac{\tau_{\theta z0}}{T_0} = \frac{\tau_{rz0}}{T_0} = 0
\end{aligned}
\right\}
\tag{2.38}
$$

23

in the core:

$$\frac{\sigma_{r1}}{T_0} = \frac{\sigma_{\theta 1}}{T_0} = -\frac{1 - \mu^2}{\mu^2} \cdot A = \text{const} \left.\vphantom{\begin{array}{c} \\ \\ \\ \\ \end{array}}\right\}$$

$$\frac{\tau_{r\theta 1}}{T_0} = \frac{\tau_{\theta z 1}}{T_0} = \frac{\tau_{rz 1}}{T_0} = 0$$

(2.39)

where

$$A = \frac{2(R + D_1)\mu^2 - D_0(N + 1)}{M\mu^2 - (1 - \mu^2)(N + 1)} - \frac{a_0 E_0}{2(1 - \nu_0)} \cdot T_0 \left.\vphantom{\begin{array}{c} \\ \\ \\ \\ \\ \\ \\ \\ \end{array}}\right\}$$

$$D_0 = [1 - \mu^2(1 - m^2)] \cdot \frac{a_0 E_0}{2(1 - \nu_0)} T_0$$

$$D_1 = m \cdot \frac{a_0 E_0}{2(1 - \nu_0)} \cdot T_0$$

$$R = 2\bar{\epsilon}_z \frac{(\nu_0 - \nu_1) G_0 G_1}{G_0 - G_1}$$

(2.40)

Here μ, m, M, N, G_0, and G_1 are the constants that have been considered previously [the notation is that of (2.6) and (2.12)].

The axial stresses in the shell and in the core are obtained from the generalized Hooke's law [relations (1.2) and (1.3)]:

$$\frac{\sigma_{zj}}{T_0} = E_j(\bar{\epsilon}_z - a_j T_0) + \nu_j \left(\frac{\sigma_{rj}}{T_0} + \frac{\sigma_{\theta j}}{T_0}\right)$$

(2.41)

$$(j = 0, 1)$$

The total axial deformation $\bar{\epsilon}_z$ of a composite cylinder is obtained from the force equilibrium along any orthogonal axis of a cylinder cross section

$$\int_0^a \frac{\sigma_{z1}}{T_0} r dr + \int_a^b \frac{\sigma_{z0}}{T_0} r dr = 0$$

(2.42)

After the substitution of the stresses $\frac{\sigma_{zj}}{T_0}$ as given by (2.41) in (2.42) and the use of (2.38) to (2.40), we obtain an equation for the constant $\bar{\epsilon}_z$ which yields the formula

$$\bar{\epsilon}_z = \frac{B(1 + \nu_0)[1 - \mu^2(1 - \gamma\delta)] + 2\mu^2(1 - \mu^2)(\nu_0 - \nu_1)[1 + \nu_0 - \gamma(1 + \nu_1)]}{B(1 + \nu_0)[1 - \mu^2(1 - \delta)] + 2\mu^2(1 - \mu^2)(\nu_0 - \nu_1)^2} \cdot a_0 E_0$$

(2.43)

where

$$B = 2(1 - \nu_0)\mu^2 + (1 - \mu^2)[1 - \delta_0 + 2\delta_0(1 - \nu_0)] \qquad \delta_0 = \frac{E_0}{E_1}$$

[see the notation in (2.6)].

24

Substituting the values of the constants R, D_0, D_1, M, N, and τ_z in the formula (2.40) for A, we obtain

$$A = a_0 E_0 T_0 \mu^2 \frac{(\nu_0 - \nu_1)[1 - \mu^2(1 - \gamma\delta)] - [1 - \mu^2(1 - \delta)][1 + \nu_0 - \gamma(1 + \nu_1)]}{B(1 + \nu_0)[1 - \mu^2(1 - \delta)] + 2\mu^2(1 - \mu^2)(\nu_0 - \nu_1)^2} \tag{2.44}$$

Using (2.38), (2.39), and (2.42) to (2.44), we write the axial deformations (2.41)

$$\left. \begin{aligned} \frac{\sigma_{z0}}{T_0} &= a_0 E_0 T_0 \mu^2(\gamma - 1) \frac{B\delta(1 + \nu_0) - 2(1 - \mu^2)(1 + \nu_1)(\nu_0 - \nu_1)}{B(1 + \nu_0)[1 - \mu^2(1 - \delta)] + 2\mu^2(1 - \mu^2)(\nu_0 - \nu_1)^2} + 2\nu_0 A \\[2ex] \frac{\sigma_{z1}}{T_0} &= -\frac{1 - \mu^2}{\mu^2} \frac{\sigma_{z0}}{T_0} \end{aligned} \right\} \tag{2.45}$$

For $T = 1$, formulas (2.38), (2.39), and (2.45) together with (2.44) determine the desired thermal stresses $\underset{1}{\sigma_{ij}}$ and $\underset{1}{\tau_{rzj}}$ $(i = r, \theta, z; j = 0, 1)$ in the shell and in the core. We clearly have

$$\frac{\sigma_{ij}}{T_0} = T_0 \cdot \underset{1}{\sigma_{ij}}$$

$$\frac{\tau_{rzj}}{T_0} = T_0 \cdot \underset{1}{\tau_{rzj}}$$

$$(i = r, \theta, z; j = 0, 1)$$

For $\nu_0 = \nu_1 = \nu$, the formulas for $\underset{1}{\sigma_{ij}}$ and $\underset{1}{\tau_{rzj}}$ take the following particularly simple forms:

in the core:

$$\left. \begin{aligned} \underset{1}{\sigma_{r1}} &= \underset{1}{\sigma_{\theta 1}} = C\delta(1 - \gamma)(1 - \mu^2)a_0 E_0 \\[2ex] \underset{1}{\sigma_{z1}} &= C\delta(1 - \gamma)(1 - \mu^2)\left(1 + \frac{\delta}{1 - \mu^2(1 - \delta)}\right)a_0 E_0 \\[2ex] \underset{1}{\tau_{rz1}} &= 0 \end{aligned} \right\} \tag{2.46}$$

in the shell:

$$\left. \begin{aligned} \underset{1}{\sigma_{r0}} &= C\delta\mu^2(\gamma - 1)\left(1 - \frac{b^2}{r^2}\right)a_0 E_0 \\[2ex] \underset{1}{\sigma_{\theta 0}} &= C\delta\mu^2(\gamma - 1)\left(1 + \frac{b^2}{r^2}\right)a_0 E_0 \\[2ex] \underset{1}{\sigma_{z0}} &= C\delta\mu^2(\gamma - 1)\left(1 + \frac{\delta}{1 - \mu^2(1 - \delta)}\right)a_0 E_0 \\[2ex] \underset{1}{\tau_{rz0}} &= 0 \end{aligned} \right\} \tag{2.47}$$

25

where

$$C = \frac{1}{2(1 - \nu) - (1 - \delta)[1 + (1 - 2\nu)\mu^2]}$$

From the results of calculations and the use of similarity theory [see (2.33)], tables were drawn up of the stresses $\overline{\sigma}_{ij}\underset{\eta_1}{}(\mu, \gamma, \delta, r, z)$ and $\overline{\tau}_{rzj}\underset{\eta_1}{}(\mu, \gamma, \delta, r, z)$ $(i = r, \theta, z; j = 0, 1)$ with an absolute error less than $|\Delta|_{\eta_1} = 0.0005$, for the parameters $\mu = a/b = 0.7(0.05)0.95$, $\gamma = a_1/a_0 = 0.5(0.5)2$, $\delta = E_1/E_0 = 0.5, 1,$ and 2, $\lambda = r/b = 0(0.1)0.7$ and $0.7(0.05)0.95$, and independent variable $p = z/\pi b = 0(0.01)0.02, 0.05(0.05)0.3,$ and $0.3(0.1)0.8$. For intermediate values of the parameters, the stresses $\overline{\sigma}_{ij}\underset{\eta_1}{}$ and $\overline{\tau}_{rzj}\underset{\eta_1}{}$ can be obtained by linear interpolation.

For $p \geq 0.8$, the stresses are independent of p and the relations (2.36) hold. These relations can be written

$$\overline{\sigma}_{ij}\underset{\eta_1}{}(\mu, \gamma, \delta, \lambda, p) = 0.5\overline{\sigma}_{ij}\underset{1}{}(\mu, \gamma, \delta, \lambda)$$

$$\overline{\tau}_{rzj}\underset{\eta_1}{}(\mu, \gamma, \delta, \lambda, p) = 0.5 \cdot \overline{\tau}_{rzj}\underset{1}{}(\mu, \gamma, \delta, \lambda)$$

$$(i = r, \theta, z; j = 0, 1)$$

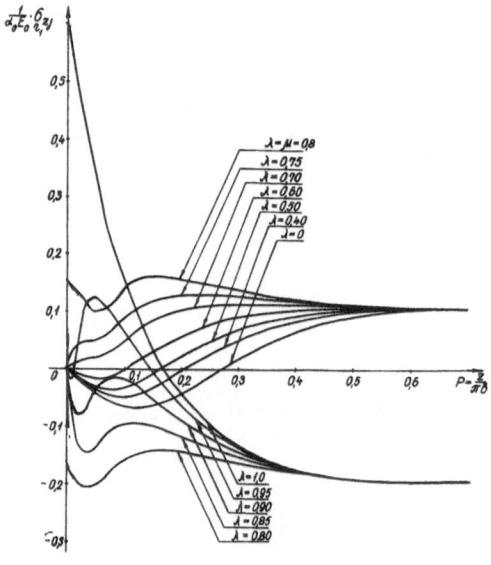

Fig. 3. The axial stresses $\sigma_{zj}\underset{\eta_1}{}$ $(j = 0, 1)$ in a composite cylinder with parameters $\mu = 0.8$, $\gamma = 0.5$, $\delta = 1$, and the temperature function $\eta_1(z)$.

26

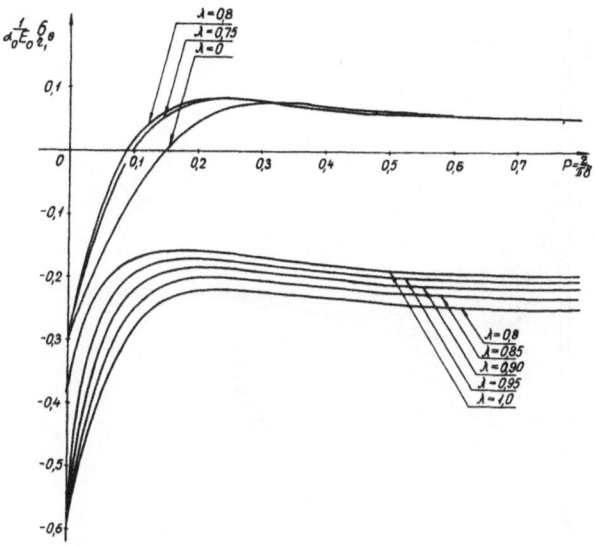

Fig. 4. The tangential stresses $\sigma_{\theta j}$ ($j = 0, 1$) in η_1 a composite cylinder with parameters $\mu = 0.8$, $\gamma = 0.5$, $\delta = 1$, and the temperature function $\eta_1(z)$.

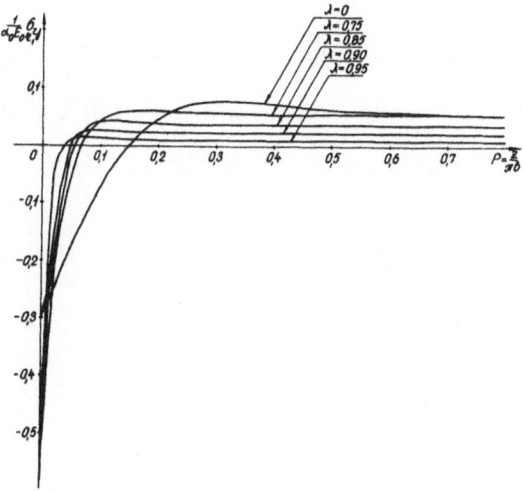

Fig. 5. The radial stresses σ_{rj} ($j = 0, 1$) in a η_1 composite cylinder with parameters $\mu = 0.8$, $\gamma = 0.5$, $\delta = 1$, and the temperature function $\eta_1(z)$.

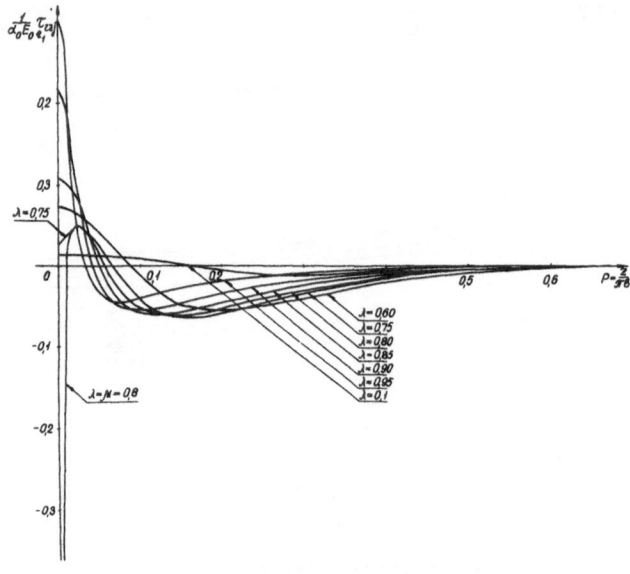

Fig. 6. The tangential stresses τ_{rzj} ($j = 0, 1$) in a composite
$\quad\quad\quad\quad\quad\quad\quad\quad\quad\quad\quad_{\eta_1}$
cylinder with parameters $\mu = 0.8$, $\gamma = 0.5$, $\delta = 1$, and the
temperature function $\eta_1(z)$.

The normal stresses are odd functions of z, the tangential stresses are even functions of z, and for negative p we have

$$\left.\begin{array}{l} \overline{\sigma}_{ij}(\mu, \gamma, \delta, \lambda, -p) = -\overline{\sigma}_{ij}(\mu, \gamma, \delta, \lambda, p) \\ {}_{\eta_1} \quad\quad\quad\quad\quad\quad\quad\quad {}_{\eta_1} \\[2mm] \overline{\tau}_{rzj}(\mu, \gamma, \delta, \lambda, -p) = \overline{\tau}_{rzj}(\mu, \gamma, \delta, \lambda, p) \\ {}_{\eta_1} \quad\quad\quad\quad\quad\quad\quad\quad {}_{\eta_1} \end{array}\right\} \quad\quad (2.48)$$

As an example of the results obtained, Figs. 3, 4, 5, and 6 show graphs of the stresses $\overline{\sigma}_{ij}$ and
$\quad\quad\quad\quad\quad\quad\quad\quad\quad\quad\quad\quad\quad\quad\quad\quad\quad\quad\quad\quad\quad\quad\quad\quad\quad\quad\quad\quad\quad\quad_{\eta_1}$
$\overline{\tau}_{rzj}$ ($i = r, \theta, z$; $j = 0, 1$) for a composite cylinder with $\mu = 0.8$, $\gamma = 0.5$, and $\delta = 1$.
${}_{\eta_1}$

28

Thermal Stresses in a Composite Cylinder with an Arbitrary Temperature Distribution Along Its Length and Constant Thermal-Expansion Coefficients in the Shell and the Core

In this chapter we assume that the thermal-expansion coefficients a_0 and a_1 of the shell and the core are independent of the temperature and so are constant. Hence the parameter $\gamma = a_1/a_0$ will also be constant.

Let the temperature of the cylinder be given by the arbitrary temperature function (0.2), specified either analytically or by a table of values. We assume that there are no external loads on the cylinder. The thermal stresses in such a cylinder must satisfy the conditions (1.5) on the interface between the shell and the core and the boundary conditions (1.6).

In the interval $[-H/2, H/2]$ we write the function (0.2) in the form

$$T(z) = T\left(-\frac{H}{2}\right) + \int_{-H/2}^{H/2} T'(\xi)\eta(z - \xi)d\xi \tag{3.1}$$

where $\eta(\xi)$ is the unit step function

$$\eta(\xi) = \begin{cases} 1 & \xi > 0 \\ 0 & \xi < 0 \end{cases} \tag{3.2}$$

By using (2.35)–(2.37), which are based on calculated results, for the temperature function (3.1) expressed in multiples of $a_0 E_0$, we can write the stresses in the shell and in the core in $-H/2 + p_0\pi b \leq z \leq H/2 - p_0\pi b$ in the form

$$\underset{T}{\overline{\sigma}_{ij}}(\mu, \gamma, \delta, r, z) = T\left(-\frac{H}{2}\right)\underset{1}{\overline{\sigma}_{ij}}(\mu, \gamma, \delta, r) + \int_{-H/2}^{H/2} T'(\xi)\underset{\eta}{\overline{\sigma}_{ij}}(\mu, \gamma, \delta, r, z - \xi)d\xi$$

$$\underset{T}{\overline{\tau}_{rzj}}(\mu, \gamma, \delta, r, z) = T\left(-\frac{H}{2}\right) \cdot \underset{1}{\overline{\tau}_{rzj}}(\mu, \gamma, \delta, r) + \int_{-H/2}^{H/2} T'(\xi)\underset{\eta}{\overline{\tau}_{rzj}}(\mu, \gamma, \delta, r, z - \xi)d\xi \tag{3.3}$$

$$(i = r, \theta, z; j = 0, 1)$$

where $\underset{\eta}{\overline{\sigma}_{ij}}$ and $\underset{\eta}{\overline{\tau}_{rzj}}$ are the stresses in a composite cylinder for the unit step function (3.2) expressed as multiples of $a_0 E_0$.

We now write the temperature function (3.2) in the form $\eta(\xi) = \eta_1(\xi) + \frac{1}{2}$, where $\eta_1(\xi)$ is the function (0.3). The stresses $\overline{\sigma}_{ij}$ and $\overline{\tau}_{rzj}$ ($i = r, \theta, z; j = 0, 1$) are

$$
\left.
\begin{array}{l}
\underset{\eta}{\overline{\sigma}_{ij}}(\mu, \gamma, \delta, r, \xi) = \underset{\eta_1}{\overline{\sigma}_{ij}}(\mu, \gamma, \delta, r, \xi) + \tfrac{1}{2}\underset{1}{\overline{\sigma}_{ij}}(\mu, \gamma, \delta, r) \\[2mm]
\underset{\eta}{\overline{\tau}_{rzj}}(\mu, \gamma, \delta, r, \xi) = \underset{\eta_1}{\overline{\tau}_{rzj}}(\mu, \gamma, \delta, r, \xi) + \tfrac{1}{2}\underset{1}{\overline{\tau}_{rzj}}(\mu, \gamma, \delta, r) \\[2mm]
(i = r, \theta, z; j = 0, 1)
\end{array}
\right\}
\tag{3.4}
$$

From (3.4) we obtain the obvious relations

$$
\left.
\begin{array}{l}
\underset{\eta}{\overline{\sigma}_{ij}}(\mu, \gamma, \delta, r, z - \xi) = \underset{\eta_1}{\overline{\sigma}_{ij}}(\mu, \gamma, \delta, r, z - \xi) + \tfrac{1}{2}\underset{1}{\overline{\sigma}_{ij}}(\mu, \gamma, \delta, r) \\[2mm]
\underset{\eta}{\overline{\tau}_{rzj}}(\mu, \gamma, \delta, r, z - \xi) = \underset{\eta_1}{\overline{\tau}_{rzj}}(\mu, \gamma, \delta, r, z - \xi) + \tfrac{1}{2}\underset{1}{\overline{\tau}_{rzj}}(\mu, \gamma, \delta, r)
\end{array}
\right\}
\tag{3.5}
$$

The substitution of (3.5) in (3.3) and the use of the equation $\underset{1}{\overline{\tau}_{rzj}}(\mu, \gamma, \delta, r) \equiv 0$ ($j = 0, 1$) yield

$$
\left.
\begin{array}{l}
\underset{T}{\overline{\sigma}_{ij}}(\mu, \gamma, \delta, r, z) = \tfrac{1}{2}\left[T\left(-\frac{H}{2}\right) + T\left(\frac{H}{2}\right)\right]\underset{1}{\overline{\sigma}_{ij}}(\mu, \gamma, \delta, r) + \int_{-H/2}^{H/2} T'(\xi)\underset{\eta_1}{\overline{\sigma}_{ij}}(\mu, \gamma, \delta, r, z - \xi)d\xi \\[4mm]
\underset{T}{\overline{\tau}_{rzj}}(\mu, \gamma, \delta, r, z) = \int_{-H/2}^{H/2} T'(\xi)\underset{\eta_1}{\overline{\tau}_{rzj}}(\mu, \gamma, \delta, r, z - \xi)d\xi \\[4mm]
(i = r, \theta, z; j = 0, 1)
\end{array}
\right\}
\tag{3.6}
$$

In (3.6) we have $\underset{\eta_1}{\overline{\sigma}_{ij}} = (1/a_0E_0)\underset{\eta_1}{\sigma_{ij}}$ and $\underset{\eta_1}{\overline{\tau}_{rzj}} = (1/a_\theta E_0)\underset{\eta_1}{\tau_{rzj}}$ ($i = r, \theta, z; j = 0, 1$) given by the series (2.8)–(2.11) and (2.23)–(2.26) and $\underset{1}{\overline{\sigma}_{ij}}(1/a_0E_0)\underset{1}{\sigma_{ij}}$ by the formulas (2.46) and (2.47). The solution of this equation yields the exact solution of the thermoelastic problem for a composite cylinder with the temperature function (3.1).

When we substitute the stresses $\underset{\eta_1}{\overline{\sigma}_{ij}}$ and $\underset{\eta_1}{\overline{\tau}_{rzj}}$ ($i = r, \theta, z; j = 0, 1$), tabulated with a precision such that $|\Delta|_{\eta_1} \leq 0.0005$, into the solution (3.6), we obtain an approximate solution of our problem with an absolute error not exceeding $a_0 E_0 \cdot |\Delta|_{\eta_1} \cdot \bigvee_{-H/2}^{H/2}(T)$, where $\bigvee_{-H/2}^{H/2}(T)$ is the total variation of the function $T(z)$ in $[-H/2, H/2]$ (this will be proved below).

Since (2.36) and (2.37) have an error $|\Delta|_{\eta_1} \leq 0.0005$, we conclude that for $\xi \leq z - 1$ ($z - \xi \geq 1$)

$$
\left.
\begin{array}{l}
\underset{\eta_1}{\overline{\sigma}_{ij}}(\mu, \gamma, \delta, r, z - \xi) = \tfrac{1}{2}\underset{1}{\overline{\sigma}_{ij}}(\mu, \gamma, \delta, r) \\[2mm]
\underset{\eta_1}{\overline{\tau}_{rzj}}(\mu, \gamma, \delta, r, z - \xi) = \tfrac{1}{2}\underset{1}{\overline{\tau}_{rzj}}(\mu, \gamma, \delta, r)
\end{array}
\right\}
\tag{3.7}
$$

and for $\xi \geq z + 1$ ($z - \xi \leq -1$)

$$\underset{\eta_1}{\overline{\sigma}_{ij}}(\mu, \gamma, \delta, r, z - \xi) = -\tfrac{1}{2}\underset{1}{\overline{\sigma}_{ij}}(\mu, \gamma, \delta, r)$$

$$\underset{\eta_1}{\overline{\tau}_{rzj}}(\mu, \gamma, \delta, r, z - \xi) = \tfrac{1}{2}\underset{1}{\overline{\tau}_{rzj}}(\mu, \gamma, \delta, r) \tag{3.8}$$

Substituting (3.7) and (3.8) into (3.6) and using the fact that $\underset{1}{\overline{\tau}_{rzj}}(\mu, \gamma, \delta, r) \equiv 0$ $(j = 0, 1)$, we obtain

$$\underset{T}{\overline{\sigma}_{ij}}(\mu, \gamma, \delta, r, z) = \tfrac{1}{2}\left[T\left(-\frac{H}{2}\right) + T\left(\frac{H}{2}\right)\right]\underset{1}{\overline{\sigma}_{ij}}(\mu, \gamma, \delta, r) + \tfrac{1}{2}\int_{-H/2}^{z-l} T'(\xi)\underset{1}{\overline{\sigma}_{ij}}(\mu, \gamma, \delta, r)d\xi$$

$$+ \int_{z-l}^{z+l} T'(\xi)\underset{\eta_1}{\overline{\sigma}_{ij}}(\mu, \gamma, \delta, r, z - \xi)d\xi - \tfrac{1}{2}\int_{z+l}^{H/2} T'(\xi)\underset{1}{\overline{\sigma}_{ij}}(\mu, \gamma, \delta, r)d\xi$$

$$= T_{av}\cdot\underset{1}{\overline{\sigma}_{ij}}(\mu, \gamma, \delta, r) + \int_{z-l}^{z+l} T'(\xi)\underset{\eta_1}{\sigma_{ij}}(\mu, \gamma, \delta, r, z - \xi)d\xi$$

$$\underset{T}{\overline{\tau}_{rzj}}(\mu, \gamma, \delta, r, z) = \int_{z-l}^{z+l} T'(\xi)\underset{\eta_1}{\overline{\tau}_{rzj}}(\mu, \gamma, \delta, r, z - \xi)d\xi$$

where $T_{av} = \tfrac{1}{2}[T(z - l) + T(z + l)]$ is the mean of the temperatures at the points $z - l$ and $z + l$, $l = p_0\pi b$, and $p_0 = 0.8$.

Making the change of variable $t = z - \xi$ in (3.6), we obtain

$$\underset{T}{\overline{\sigma}_{ij}}(\mu, \gamma, \delta, r, z) = T_{av}\cdot\underset{1}{\overline{\sigma}_{ij}}(\mu, \gamma, \delta, r) + \int_{-l}^{l} T'(z - t)\underset{\eta_1}{\overline{\sigma}_{ij}}(\mu, \gamma, \delta, r, t)dt$$

$$\underset{T}{\overline{\tau}_{rzj}}(\mu, \gamma, \delta, r, z) = \int_{-l}^{l} T'(\xi)\underset{\eta_1}{\overline{\tau}_{rzj}}(\mu, \gamma, \delta, r, t)dt \tag{3.9}$$

$$(i = r, \theta, z; \ j = 0, 1)$$

Using the relations (2.33), which were obtained by applying similarity theory, we calculated the stresses $\underset{\eta_1}{\overline{\sigma}_{ij}}(\mu, \gamma, \delta, r, z)$ and $\underset{\eta_1}{\overline{\tau}_{rzj}}(\mu, \gamma, \delta, r, z)$ $(i = r, \theta, z; \ j = 0, 1)$ and tabulated the results for various values of the parameters $\mu, \gamma, \delta, \lambda = r/b$, and the independent variable $p = z/\pi b$, so that

$$\underset{\eta_1}{\overline{\sigma}_{ij}}(\mu, \gamma, \delta, r, z) = \underset{\eta_1}{\overline{\sigma}_{ij}}(\mu, \gamma, \delta, \lambda, p)$$

$$\underset{\eta_1}{\overline{\tau}_{rzj}}(\mu, \gamma, \delta, r, z) = \underset{\eta_1}{\overline{\tau}_{rzj}}(\mu, \gamma, \delta, \lambda, p) \tag{3.10}$$

If we use the relations (3.10) and make the change of variable $t/\pi b = -p$ in (3.9), then by using (2.48)

31

we obtain

$$
\underset{T}{\overline{\sigma}_{ij}}(\mu,\,\gamma,\,\delta,\,r,\,z) = T_{av}\cdot\underset{1}{\overline{\sigma}_{ij}}(\mu,\,\gamma,\,\delta,\,\lambda) + \int_{-l}^{l} T'(z-t)\underset{\eta_1}{\overline{\sigma}_{ij}}\left(\mu,\,\gamma,\,\delta,\,\frac{r}{b},\,\frac{t}{\pi b}\right)dt
$$

$$
= T_{av}\cdot\underset{1}{\overline{\sigma}_{ij}}(\mu,\,\gamma,\,\delta,\,\lambda) - \pi b\int_{-P_0}^{P_0} T'(z+\pi bp)\underset{\eta_1}{\overline{\sigma}_{ij}}(\mu,\,\gamma,\,\delta,\,\lambda,\,p)dp
$$

$$
\underset{T}{\overline{\tau}_{rzj}}(\mu,\,\gamma,\,\delta,\,r,\,z) = \pi b\cdot\int_{-P_0}^{P_0} T'(z+\pi bp)\underset{\eta_1}{\overline{\tau}_{rzj}}(\mu,\,\gamma,\,\delta,\,\lambda,\,p)dp
$$

$$
(i = r,\,\theta,\,z;\ j = 0,\,1)
$$

(3.11)

or, in terms of Stieltjes integrals,

$$
\underset{T}{\overline{\sigma}_{ij}}(\mu,\,\gamma,\,\delta,\,r,\,z) = T_{av}\cdot\underset{1}{\overline{\sigma}_{ij}}(\mu,\,\gamma,\,\delta,\,\lambda) - \int_{-P_0}^{P_0}\underset{\eta_1}{\overline{\sigma}_{ij}}(\mu,\,\gamma,\,\delta,\,\lambda,\,p)dT(z+\pi bp)
$$

$$
\underset{T}{\overline{\tau}_{rzj}}(\mu,\,\gamma,\,\delta,\,r,\,z) = \int_{-P_0}^{P_0}\underset{\eta_1}{\overline{\tau}_{rzj}}(\mu,\,\gamma,\,\delta,\,\lambda,\,p)dT(z+\pi bp)
$$

$$
(i = r,\,\theta,\,z;\ j = 0,\,1)
$$

(3.12)

The stresses in a composite cylinder for the temperature function (3.1), expressed in terms of the stresses $\underset{T}{\overline{\sigma}_{ij}}$ and $\underset{T}{\overline{\tau}_{rzj}}$ $(i = r,\,\theta,\,z;\ j = 0,\,1)$, are

$$
\underset{T}{\sigma_{ij}} = a_0 E_0 \underset{T}{\overline{\sigma}_{ij}}
$$

$$
\underset{T}{\tau_{rzj}} = a_0 E_0 \underset{T}{\overline{\tau}_{rzj}}
$$

$$
(i = r,\,\theta,\,z;\ j = 0,\,1)
$$

(3.13)

The displacements in a composite cylinder for the temperature function (3.1) can be obtained similarly, and are given by

$$
\underset{T}{u_j} = a_0\cdot\left\{0.5\left[T\left(-\frac{H}{2}\right) + T\left(\frac{H}{2}\right)\right]\underset{1}{\overline{u}_j}(\mu,\,\gamma,\,\delta,\,r) + \int_{-H/2}^{H/2} T'(\xi)\underset{\eta_1}{\overline{u}_j}(\mu,\,\gamma,\,\delta,\,r,\,z-\xi)d\xi\right\}
$$

$$
\underset{T}{w_j} = a_0\cdot\left\{0.5\left[T\left(-\frac{H}{2}\right) + T\left(\frac{H}{2}\right)\right]\underset{1}{\overline{w}_j}(\mu,\,\gamma,\,\delta,\,z) + \int_{-H/2}^{H/2} T'(\xi)\underset{\eta_1}{\overline{w}_j}(\mu,\,\gamma,\,\delta,\,r,\,z-\xi)d\xi\right\}
$$

where $\underset{1}{\overline{u}_j}$ and $\underset{1}{\overline{w}_j}$ $(j = 0,\,1)$ are the radial and axial displacements as multiples of a_0 in the shell and the core of a composite cylinder for the temperature function $T = 1$; this function is a special case of the function $T = T(r)$ for which the thermoelastic problem is assumed to be plane; $\underset{\eta_1}{\overline{u}_j}$ and $\underset{\eta_1}{\overline{w}_j}$ $(j = 0,\,1)$

32

are the radial and axial displacements, expressed as multiples of a_0, in the shell and core with the temperature function $\eta_1(z)$, obtained from the series (2.2) and (2.3).

The solution of the problem of the thermal stresses in a composite cylinder with an arbitrary temperature distribution that is a function of the coordinate z measured along the length of the cylinder is valid for points whose distances from the ends are at least $l = p_0\pi b$ for cylinders of length $H \geq 2l$.

The thermal stresses $\underset{T}{\sigma_{ij}}$ and $\underset{T}{\tau_{rzj}}$ ($i = r, \theta, z; j = 0, 1$) and the displacements $\underset{T}{u_j}$ and $\underset{T}{w_j}$ ($j = 0, 1$) in a composite cylinder, for the arbitrary temperature function (3.1), satisfy conditions (1.5) on the interface between the core and the shell as well as the boundary conditions (1.6). In fact, on the interface ($r = a$, $\lambda = \mu = a/b$) we have

$$\underset{T}{u_0} = \underset{T}{u_1} \qquad \underset{T}{w_0} = \underset{T}{w_1} \qquad \underset{T}{\sigma_{r0}} = \underset{T}{\sigma_{r1}} \qquad \underset{T}{\tau_{rz0}} = \underset{T}{\tau_{rz1}}$$

since for $\lambda = \mu$: $\underset{\eta_1}{u_0} = \underset{\eta_1}{u_1}$, $\underset{\eta_1}{w_0} = \underset{\eta_1}{w_1}$, $\underset{1}{u_0} = \underset{1}{u_1}$, $\underset{1}{w_0} = \underset{1}{w_1}$, $\underset{1}{\sigma_{r0}} = \underset{1}{\sigma_{r1}}$, $\underset{\eta_1}{\tau_{rz0}} = \underset{\eta_1}{\tau_{rz1}}$, $\underset{1}{\sigma_{r0}} = \underset{1}{\sigma_{r1}}$, and $\underset{1}{\tau_{rz0}} = \underset{1}{\tau_{rz1}}$. On the surface of a composite cylinder ($r = b$, $\lambda = 1$) we have $\underset{T}{\sigma_{r0}} = \underset{T}{\tau_{rz0}} = 0$, since for $\lambda = 1$

$$\underset{\eta_1}{\sigma_{r0}} = \underset{\eta_1}{\tau_{rz0}} = \underset{1}{\sigma_{r0}} = \underset{1}{\tau_{rz0}} = 0$$

The axial stresses $\underset{T}{\sigma_{z0}}$ and $\underset{T}{\sigma_{z1}}$ on any orthogonal axis of a cross section of the cylinder are in equilibrium for $|z| \leq H/2 - l$, as can be seen from (2.7) and (2.42). The tangential stresses in these sections are in equilibrium from the symmetry conditions.

The stresses $\underset{\eta_1}{\sigma_{ij}}(\mu, \gamma, \delta, \lambda, p)$ and $\underset{\eta_1}{\tau_{rzj}}(\mu, \gamma, \delta, \lambda, p)$ ($i = r, \theta, z; j = 0, 1$) obtained in the form of tables (or graphs) can, with sufficient accuracy for practical purposes, be approximated by elementary functions, and the integrals (3.11) can be expressed in analytic, finite form for many practically important cases.

When it is difficult to obtain finite expressions for the integrals (3.11), the numerical methods developed in Chapter 5 may be used.

Estimate of the Error in the Approximate Solution

To shorten the calculations, we estimate the errors arising in the calculation of the normal stresses from the approximate formulas (3.11) or (3.12) obtained from (3.6). All these calculations are also valid for tangential stresses. We denote the approximate normal stresses in a composite cylinder with the temperature function $T(z)$ by $\underset{T}{\sigma_{ij}^*}$ ($i = r, \theta, z; j = 0, 1$), and those in one with the temperature function $\eta_1(z)$ by $\underset{\eta_1}{\sigma_{ij}^*}$.[†]

The differences between exact stresses and approximate stresses obtained from formulas (3.6) with the use of (3.13) can be expressed by Stieltjes integrals:

$$\underset{T}{\sigma_{ij}}(\mu, \gamma, \delta, r, z) - \underset{T}{\sigma_{ij}^*}(\mu, \gamma, \delta, r, z) = a_0 E_0 \int_{-H/2}^{+H/2} \left[\underset{\eta_1}{\overline{\sigma}_{ij}}(\mu, \gamma, \delta, r, z - \xi) - \underset{\eta_1}{\overline{\sigma}_{ij}^*}(\mu, \gamma, \delta, r, z - \xi) \right] dT(\xi)$$

[†]The asterisk is used to denote approximate stresses only in this estimate of the accuracy of solutions. It is not used elsewhere.

for which we have the following inequalities [9]:

$$\left|\sigma_{ij} - \sigma_{ij}^*\right| = a_0 E_0 \left|\int_{-H/2}^{H/2}\left(\overline{\sigma}_{ij} - \overline{\sigma}_{ij}^*\right)dT(\xi)\right| \leq \max\left|\overline{\sigma}_{ij} - \overline{\sigma}_{ij}^*\right| \cdot \bigvee_{-H/2}^{H/2}(T)a_0 E_0$$

where $\displaystyle\bigvee_{-H/2}^{H/2}(T)$ is the total variation of the temperature function $T(z)$ in the interval $[-H/2, H/2]$. If

$T(z)$ is monotonic, $\displaystyle\bigvee_{-H/2}^{H/2}(T) = |T(H/2) - T(-H/2)|$.

The stresses $\overline{\sigma}_{ij}^*$ were calculated and tabulated with an absolute error $|\Delta|_{\eta_1} < 0.0005$ for all η_1

values of r and z and of the parameters μ, γ, and δ. This means that for all values of the independent variables r and z and of the parameters μ, γ, and δ we have

$$\max\left|\overline{\sigma}_{ij} - \overline{\sigma}_{ij}^*\right| \leq |\Delta|_{\eta_1}$$
$$\quad\;\; \eta_1 \quad \eta_1$$

Using this inequality, we obtain a final estimate of the error of the approximate solution:

$$|\Delta|_T = \left|\sigma_{ij}(\mu, \gamma, \delta, r, z) - \sigma_{ij}^*(\mu, \gamma, \delta, r, z)\right| \leq a_0 E_0 |\Delta|_{\eta_1} \cdot \bigvee_{-H/2}^{H/2}(T) \qquad (3.14)$$

The first term in (3.6), $\frac{1}{2}[T(H/2) - T(-H/2)]\overline{\sigma}_{ij}$, is the same for both the exact and the approximate so-

lutions and thus disappears in the difference between these solutions. In the calculation of the stresses σ_{ij} from formulas (2.46) and (2.47) with an error $|\Delta|_{\eta_1}$, the best accuracy is obtained by using formula

(3.14) for calculating the σ_{ij}^*. The inequality (3.14) gives the error in the approximate solution (3.11)

or (3.12) due to the replacement of the exact stresses σ_{ij} and τ_{rzj} by their approximate values σ_{ij}^* and η_1

τ_{rzj}^* with an error of $|\Delta|_{\eta_1}$.

If the temperature function has an arbitrary discontinuity

$$T(z) = \begin{cases} T_1 & -\dfrac{H}{2} \leq z < z_0 \\[2em] T_2 & z_0 < z \leq \dfrac{H}{2} \end{cases}$$

where T_1 and T_2 are arbitrary constants and z_0 satisfies (2.34), it can be written as the sum

$$T(z) = \frac{1}{2}(T_1 + T_2) + (T_2 - T_1)\eta_1(z - z_0)$$

From (2.35)–(2.37) the approximate thermal stresses for an arbitrary temperature discontinuity are

$$
\sigma_{ij}(\mu, \gamma, \delta, r, z) = \begin{cases}
T_1 \underset{1}{\sigma_{ij}}(\mu, \gamma, \delta, \lambda) & -\dfrac{H}{2} \le z \le z_0 - l \\[2ex]
\tfrac{1}{2}(T_1 + T_2)\underset{1}{\sigma_{ij}}(\mu, \gamma, \delta, \lambda) + (T_2 - T_1)\underset{\eta_1}{\sigma_{ij}}\left(\mu, \gamma, \delta, \lambda, p - \dfrac{z_0}{\pi b}\right) \\[2ex]
\qquad\qquad\qquad\qquad\qquad\qquad\qquad\qquad z_0 - l \le z \le z_0 + l \\[2ex]
T_2 \underset{1}{\sigma_{ij}}(\mu, \gamma, \delta, \lambda) & z_0 + l \le z \le \dfrac{H}{2}
\end{cases}
$$

$$
\underset{T}{\tau_{rzj}}(\mu, \gamma, \delta, r, z) = \begin{cases}
(T_2 - T_1)\underset{\eta_1}{\tau_{rzj}}\left(\mu, \gamma, \delta, \lambda, p - \dfrac{z_0}{\pi b}\right) & z_0 - l \le z \le z_0 + l \\[2ex]
0 & -\dfrac{H}{2} \le z \le z_0 - l, \; z_0 + l \le z \le \dfrac{H}{2}
\end{cases}
$$

$$
(i = r, \theta, z; \; j = 0, 1)
$$

The thermal stresses in a composite cylinder are strongly affected by the initial temperature distribution at which the cylinder is formed in its natural, unstressed state, and from which the temperatures are calculated.

For practical calculations in the case under consideration, the temperature function (0.2) can be written

$$
T(z) = t(z) - t_0(z) \tag{3.15}
$$

where $t(z)$ is the current temperature distribution in the composite cylinder and $t_0(z)$ the initial temperature distribution. Note that the origin for the temperature function (3.15) can be taken at any point and need not be the midpoint of the cylinder.

Example 1. We will calculate the thermal stresses in a composite cylinder for the linear temperature function

$$
T(z) = Az + B \qquad (A, B = \text{const}) \tag{3.16}
$$

Substituting (3.16) in (3.11), we obtain

$$
\underset{T}{\overline{\sigma}_{ij}}(\mu, \gamma, \delta, r, z) = T(z) \cdot \underset{1}{\overline{\sigma}_{ij}}(\mu, \gamma, \delta, \lambda) \qquad (i = r, \theta, z; \; j = 0, 1)
$$

or

$$
\underset{T}{\sigma_{ij}}(\mu, \gamma, \delta, r, z) = T(z)\underset{1}{\sigma_{ij}}(\mu, \gamma, \delta, \lambda)
$$

since, because the stresses $\underset{\eta_1}{\overline{\sigma}_{ij}}(\mu, \gamma, \delta, \lambda, p)$ are odd functions of p, we have

$$
\int_{-p_0}^{p_0} T'(z + \pi b p)\underset{\eta_1}{\overline{\sigma}_{ij}}(\mu, \gamma, \delta, \lambda, p)dp = A \int_{-p_0}^{p_0} \underset{\eta_1}{\overline{\sigma}_{ij}}(\mu, \gamma, \delta, \lambda, p)dp = 0
$$

and

$$T_{\mathrm{av}} = \tfrac{1}{2}[T(z-1) + T(z+1)] = Az + B = T(z)$$

The stresses $\sigma_{1j}(\mu, \gamma, \delta, \lambda)$ are given by (2.46) and (2.47).

Taking into consideration the symmetry of the functions τ_{rzj} $(j = 0, 1)$, according to (3.11), we obtain the shearing stress at temperature (3.16) in the form

$$\underset{T}{\tau_{rzj}}(\mu, \gamma, \delta, r, z) = A \int_{-p}^{p_0} \underset{\eta_1}{\tau_{rzj}}(\mu, \gamma, \delta, \lambda, p)dp = 2A \int_0^{p_0} \underset{\eta_1}{\tau_{rzj}}(\mu, \gamma, \delta, \lambda, p)dp$$

It is easily shown that, for any value of r and z, the boundary conditions (1.5) yield (see Chapter 2)

$$\int_0^r \underset{\eta_1}{\sigma_{z1}} r dr = -r \int_0^z \underset{\eta_1}{\tau_{rz1}} dz \qquad 0 \leq r \leq a$$

$$\int_0^a \underset{\eta_1}{\sigma_{z1}} r dr + \int_a^r \underset{\eta_1}{\sigma_{z0}} r dr = -r \int_0^z \underset{\eta_1}{\tau_{rz0}} dz \qquad a \leq r \leq b$$

which for $\lambda = r/b$ and $p = z/\pi b$ become

$$\int_0^\lambda \underset{\eta_1}{\sigma_{z1}} \lambda d\lambda = -\lambda \pi \int_0^p \underset{\eta_1}{\tau_{rz1}} dp \qquad 0 \leq \lambda \leq \mu$$

$$\int_0^\mu \underset{\eta_1}{\sigma_{z1}} \lambda d\lambda + \int_\mu^\lambda \underset{\eta_1}{\sigma_{z0}} \lambda d\lambda = -\lambda \pi \int_0^p \underset{\eta_1}{\tau_{rz0}} dp \qquad \mu \leq \lambda \leq 1$$

From these equations, taking into account (2.36), (2.46), and (2.47), we obtain

$$\int_0^{p_0} \underset{\eta_1}{\tau_{rz1}} dp = -\frac{1}{\lambda\pi} \int_0^\lambda \left(\underset{\eta_1}{\sigma_{z1}}\right)_{p=p_0} \lambda d\lambda = -\frac{\lambda}{4\pi} \underset{1}{\sigma_{z1}} = -\frac{\lambda}{4\pi}\left(1 - \frac{1}{\mu^2}\right)\underset{1}{\sigma_{z0}} \qquad (0 \leq \lambda \leq \mu)$$

$$\int_0^{p_0} \underset{\eta_1}{\tau_{rz0}} dp = -\frac{1}{\lambda\pi}\left[\int_0^\mu \left(\underset{\eta_1}{\sigma_{z1}}\right)_{p=p_0} \lambda d\lambda + \int_\mu^\lambda \left(\underset{\eta_1}{\sigma_{z0}}\right)_{p=p_0} \lambda d\lambda\right] = -\frac{1}{2\pi\lambda}\left[\frac{\mu^2}{2}\underset{1}{\sigma_{z1}} + \frac{\lambda^2 - \mu^2}{2}\underset{1}{\sigma_{z0}}\right]$$

$$= \frac{1 - \lambda^2}{4\pi\lambda}\underset{1}{\sigma_{z0}} \qquad (\mu \leq \lambda \leq 1)$$

and so

$$\underset{T}{\tau_{rz1}} = -\frac{A\lambda}{2\pi}\left(1 - \frac{1}{\mu^2}\right)\underset{1}{\sigma_{z0}}, \quad \underset{T}{\tau_{rz0}} = \frac{A(1 - \lambda^2)}{2\pi\lambda}\underset{1}{\sigma_{z0}}$$

Example 2. We calculate the thermal stresses in a composite cylinder for the temperature function (see Fig. 7)

$$
T(z) = \begin{cases}
T_1 & -\dfrac{H}{2} \leq z \leq z_0 \\[3mm]
\dfrac{\Delta T}{L}(z - z_0) + T_1 & z_0 \leq z \leq z_0 + L \\[3mm]
T_2 & z_0 + L \leq z \leq \dfrac{H}{2}
\end{cases} \tag{3.17}
$$

where $\Delta T = T_2 - T_1$, T_1 and T_2 being arbitrary constants, and z_0 is the coordinate of an arbitrary point of the cylinder.

Temperature distributions (3.17) or distributions close to this type are encountered in practice. We substitute (3.17) in (3.11) and obtain the required stresses.

We give the stresses for the temperature function (3.17) for three possible cases.

1. $L \geq 2l$, $l = \pi b p_0$.

$$
\underset{T}{\sigma_{ij}}(\mu, \gamma, \delta, r, z) = \begin{cases}
T_1 \underset{1}{\sigma_{ij}}(\mu, \gamma, \delta, \lambda) & -\dfrac{H}{2} \leq z \leq z_0 - l \\[4mm]
\left[\dfrac{\Delta T}{2L}(z + l - z_0) + T_1\right]\underset{1}{\sigma_{ij}}(\mu, \gamma, \delta, \lambda) - \dfrac{\pi b \Delta T}{L}\displaystyle\int\limits_{|z_0/\pi b - p|}^{p_0} \underset{\eta_1}{\sigma_{ij}}(\mu, \gamma, \delta, \lambda, p)\, dp \\[6mm]
\qquad\qquad\qquad\qquad\qquad\qquad z_0 - l \leq z \leq z_0 + l \\[3mm]
T(z)\underset{1}{\sigma_{ij}}(\mu, \gamma, \delta, \lambda) & z_0 + l \leq z \leq z_0 + L + l \\[4mm]
\tfrac{1}{2}\left[\dfrac{\Delta T}{L}(z - l - z_0) + T_1 + T_2\right]\underset{1}{\sigma_{ij}}(\mu, \gamma, \delta, \lambda) \\[4mm]
\qquad + \dfrac{\pi b \Delta T}{L}\displaystyle\int\limits_{|(z_0+L)/\pi b - p|}^{p_0} \underset{\eta_1}{\sigma_{ij}}(\mu, \gamma, \delta, \lambda, p)\, dp \\[6mm]
\qquad\qquad\qquad\qquad\qquad z_0 + L - l \leq z \leq z_0 + L + l \\[3mm]
T_2 \underset{1}{\sigma_{ij}}(\mu, \gamma, \delta, \lambda) & z_0 + L + l \leq z \leq \dfrac{H}{2}
\end{cases} \tag{3.18}
$$

2. $1 \le L \le 2l$.

$$
\underset{T}{\sigma_{ij}}(\mu, \gamma, \delta, r, z) = \begin{cases}
\underset{1}{T_1 \sigma_{ij}}(\mu, \gamma, \delta, \lambda) \qquad -\dfrac{H}{2} \le z \le z_0 - l \\[2em]
\left[\dfrac{\Delta T}{2L}(z + l - z_0) + T_1\right]\underset{1}{\sigma_{ij}}(\mu, \gamma, \delta, \lambda) - \dfrac{\pi b \Delta T}{L} \int\limits_{|z_0/\pi b - p|}^{p_0} \underset{\eta_1}{\sigma_{ij}}(\mu, \gamma, \delta, \lambda, p)\,dp \\[1em]
\qquad\qquad\qquad\qquad z_0 - l \le z \le z_0 + L - l \\[2em]
\tfrac{1}{2}(T_1 + T_2)\underset{1}{\sigma_{ij}}(\mu, \gamma, \delta, \lambda) + \dfrac{\pi b \Delta T}{L} \int\limits_{(z_0+L)/\pi b - p}^{p_0 - z_0/\pi b} \underset{\eta_1}{\sigma_{ij}}(\mu, \gamma, \delta, \lambda, p)\,dp \\[1em]
\qquad\qquad\qquad\qquad z_0 + L - l \le z \le z_0 + l \\[2em]
\tfrac{1}{2}\left[\dfrac{\Delta T}{L}(z - l - z_0) + T_1 + T_2\right]\underset{1}{\sigma_{ij}}(\mu, \gamma, \delta, \lambda) \\[1em]
\qquad + \dfrac{\pi b \Delta T}{L} \int\limits_{|(z_0+L)/\pi b - p|}^{p_0} \underset{\eta_1}{\sigma_{ij}}(\mu, \gamma, \delta, \lambda, p)\,dp \\[1em]
\qquad\qquad\qquad\qquad z_0 + l \le z \le z_0 + L + l \\[2em]
\underset{1}{T_2 \sigma_{ij}}(\mu, \gamma, \delta, \lambda) \qquad z_0 + L + l \le z \le \dfrac{H}{2}
\end{cases}
\tag{3.19}
$$

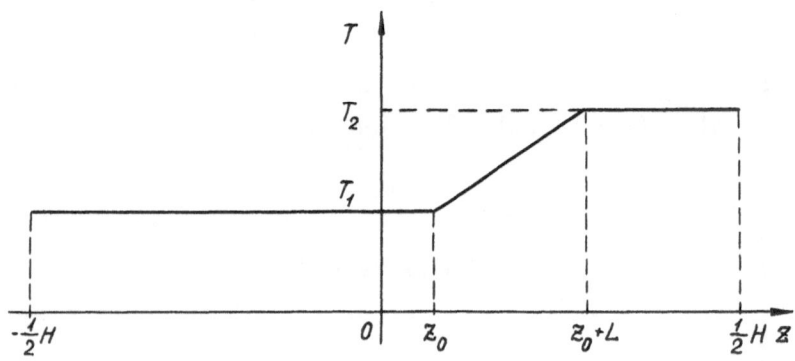

Fig. 7

38

3. $L \leq l$.

$$\sigma_{ij}(\mu, \gamma, \delta, r, z) = \begin{cases}
T_1 \cdot \underset{1}{\sigma_{ij}}(\mu, \gamma, \delta, \lambda, p) \qquad -\dfrac{H}{2} \leq z \leq z_0 - l \\[3em]
\left[\dfrac{\Delta T}{2L}(z + l - z_0) + T_1\right]\underset{1}{\sigma_{ij}}(\mu, \gamma, \delta, \lambda) - \dfrac{\pi b \Delta T}{L} \displaystyle\int\limits_{z_0/\pi b - p}^{p_0} \underset{\eta_1}{\sigma_{ij}}(\mu, \gamma, \delta, \lambda, p)\,dp \\[1em]
\hspace{12em} z_0 - l \leq z \leq z_0 + L - l \\[3em]
\tfrac{1}{2}(T_1 + T_2)\underset{1}{\sigma_{ij}}(\mu, \gamma, \delta, \lambda) - \dfrac{\pi b \Delta T}{L} \displaystyle\int\limits_{z_0/\pi b - p}^{(z_0+L)/\pi b - p} \underset{\eta_1}{\sigma_{ij}}(\mu, \gamma, \delta, \lambda, p)\,dp \\[1em]
\hspace{12em} z_0 + L - l \leq z \leq z_0 \\[3em]
\tfrac{1}{2}(T_1 + T_2)\underset{1}{\sigma_{ij}}(\mu, \gamma, \delta, \lambda) + \dfrac{\pi b \Delta T}{L} \displaystyle\int\limits_{|p-(z_0+L)/\pi b|}^{p - z_0/\pi b} \underset{\eta_1}{\sigma_{ij}}(\mu, \gamma, \delta, \lambda, p)\,dp \qquad (3.20) \\[1em]
\hspace{12em} z_0 \leq z \leq z_0 + l \\[3em]
\tfrac{1}{2}\left[\dfrac{\Delta T}{L}(z - l - z_0) + T_1 + T_2\right]\underset{1}{\sigma_{ij}}(\mu, \gamma, \delta, \lambda) \\[2em]
\qquad + \dfrac{\pi b \Delta T}{L} \displaystyle\int\limits_{|p-(z_0+L)/\pi b|}^{p_0} \underset{\eta_1}{\sigma_{ij}}(\mu, \gamma, \delta, \lambda, p)\,dp \\[1em]
\hspace{12em} z_0 + L \leq z \leq z_0 + L + l \\[3em]
T_2 \underset{1}{\sigma_{ij}}(\mu, \gamma, \delta, \lambda) \qquad z_0 + L + l \leq z \leq \dfrac{H}{2}
\end{cases}$$

The integrals in (3.18)–(3.20) can easily be evaluated from values of $\underset{\eta_1}{\sigma_{ij}}(\mu, \gamma, \delta, \lambda, p)$ given in tables or graphs.

We will calculate the thermal stresses on the surface of a composite cylinder ($r = b$, $\lambda = 1$) with the parameters $\mu = 0.8$, $\gamma = 0.5$, and $\delta = 1$. Graphs of the stresses $\underset{\eta_1}{\sigma_{ij}}$ for these parameters are shown in Figs. 3, 4, 5, and 6. On the surface of the cylinder $\underset{T}{\sigma_{r0}}$ and $\underset{T}{\tau_{rz0}}$ are zero. We calculate the stresses $\underset{T}{\sigma_{z0}}$ and $\underset{T}{\sigma_{\theta 0}}$ for $T_1 = -10°C$ and $T_2 = -40°C$ for the four lengths $L = 3l$, $L = 2l$, $L = l$, and $L = \tfrac{1}{2}l$.

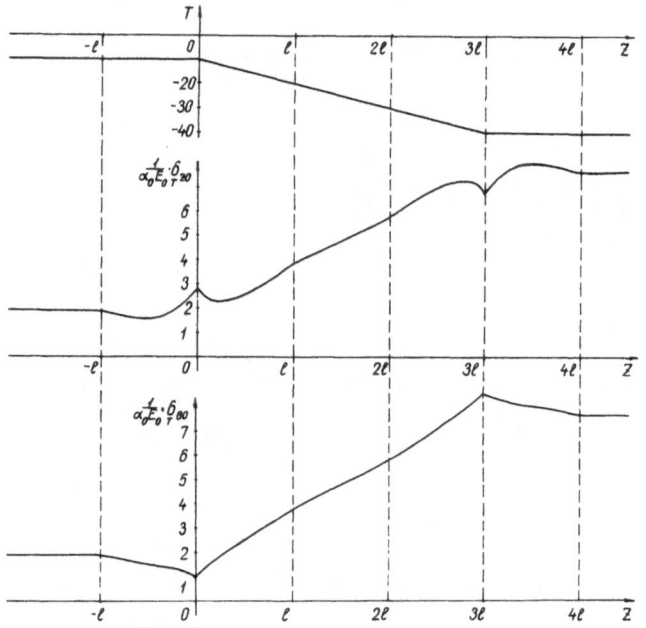

Fig. 8. Thermal stresses on the surface of a composite cylinder with the parameters $\mu = 0.8$, $\gamma = 0.5$, and $\delta = 1$, for the temperature function (3.20) and $L = 3l$.

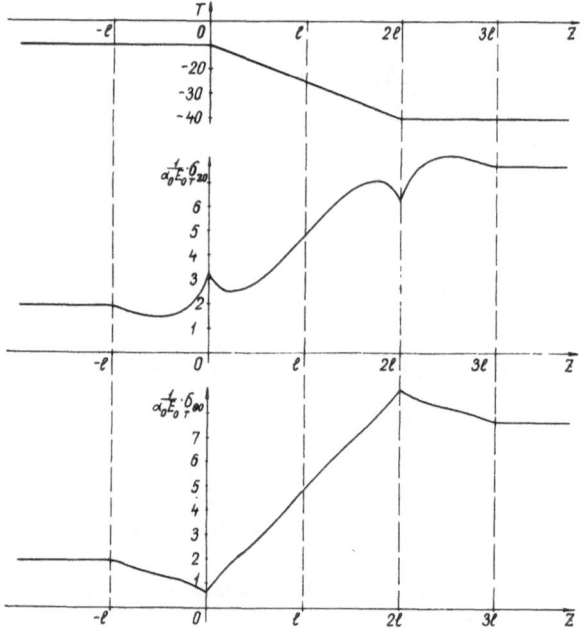

Fig. 9. Thermal stresses on the surface of a composite cylinder with the parameters $\mu = 0.8$, $\gamma = 0.5$, and $\delta = 1$, for the temperature function (3.20) and $L = 2l$.

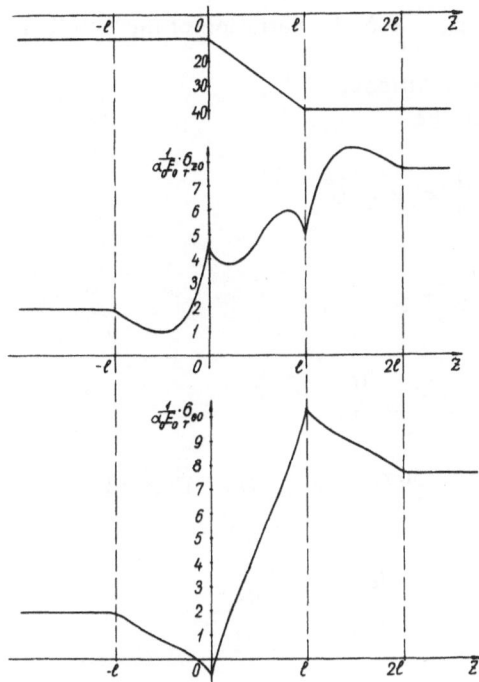

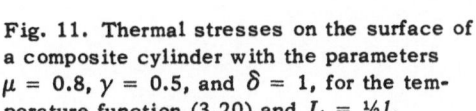

Fig. 10. Thermal stresses on the surface of a composite cylinder with the parameters $\mu = 0.8$, $\gamma = 0.5$, and $\delta = 1$, for the temperature function (3.20) and $L = l$.

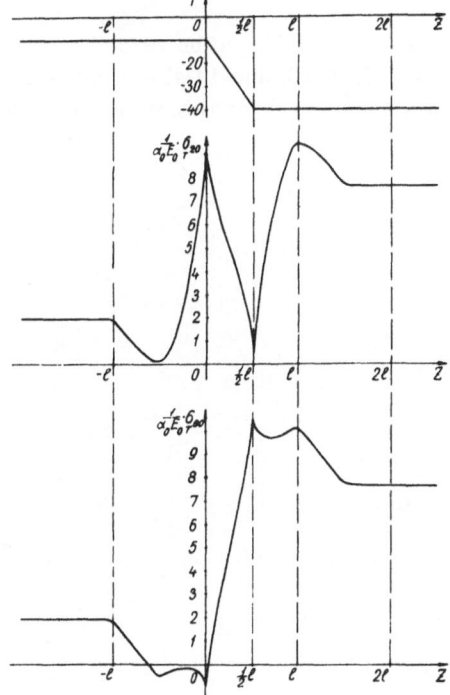

Fig. 11. Thermal stresses on the surface of a composite cylinder with the parameters $\mu = 0.8$, $\gamma = 0.5$, and $\delta = 1$, for the temperature function (3.20) and $L = \frac{1}{2}l$.

It is known that $l = p_0 \pi b$, where p_0 is chosen to satisfy the condition that σ_{ij}/η_1 be independent of p for $p \geq p_0$. In our case, we can use $p_0 = 0.6$, since for $p_0 \geq 0.6$ the variation of the function σ_{ij}/η_1 is less than 0.005, and this can be neglected in practical calculations.

A concrete temperature distribution in this example could be

$$
t(z) = \begin{cases} +5°C & -\dfrac{H}{2} \leq z \leq 0 \\[2mm] -\dfrac{30}{L} z°C & 0 \leq z \leq L \\[2mm] -25°C & L \leq z \leq \dfrac{H}{2} \end{cases}
$$

and the initial temperature distribution could be $t_0(z) = 15°C$, $-H/2 \leq z \leq H/2$. From (3.15) the calculated temperature distribution in the cylinder is

$$
T(z) = t(z) - t_0(z) = \begin{cases} -10°C & -\dfrac{H}{2} \leq z \leq 0 \\[2mm] -\dfrac{30}{L} z°C & 0 \leq z \leq L \\[2mm] -40°C & L \leq z \leq \dfrac{H}{2} \end{cases} \tag{3.21}
$$

In this case z_0 is taken to be zero.

Figures 8, 9, 10, and 11 show graphs of the temperature functions (3.21) for various values of L and of the corresponding thermal stresses σ_{z0}^{T} and $\sigma_{\theta 0}^{T}$ on the surface of a composite cylinder.

CHAPTER 4

Thermal Stresses in a Composite Cylinder with Different Thermal-Expansion Coefficients in the Shell and the Core and an Arbitrary Temperature Distribution Along Its Length

Let the thermal-expansion coefficients a_0 and a_1 of the shell and the core depend on the temperature, and let them satisfy the condition (0.5)

$$\frac{a_1(T)}{a_0(T)} = \gamma = \text{const}$$

The thermal-expansion coefficients $a_i(T)$ $(i = 0, 1)$ and the temperature function $T(z)$ occur in the original differential equations (1.1) and in the expressions for the stresses (1.4) in the form of their product: in the shell $a_0(T)T(z) = \overline{a}_0(z)T(z)$, and in the core $a_1(T)T(z) = \gamma a_0(T)T(z) = \gamma\overline{a}_0(z)T(z)$.

We call the function

$$u(z) = \overline{a}_0(z) \cdot T(z) \tag{4.1}$$

the generalized temperature function; for this function the thermal-expansion coefficient in the shell is $a_0 = 1$ and that in the core is $a_1 = \gamma$.

We obtain the thermal stresses in a composite cylinder for the temperature function (0.2) and for the thermal-expansion coefficients (0.3) by replacing $T(z)$ by the function (4.1) and setting $a_0 = 1$:

$$\underset{T}{\sigma_{ij}}(\mu, \gamma, \delta, r, z) = E_0\left[u_{av} \cdot \underset{1}{\overline{\sigma}_{ij}}(\mu, \gamma, \delta, \lambda) - \pi b \int_{-p_0}^{p_0} u'(z + \pi b p)\underset{\eta_1}{\overline{\sigma}_{ij}}(\mu, \gamma, \delta, \lambda, p)\,dp\right]$$

$$\underset{T}{\tau_{rzj}}(\mu, \gamma, \delta, r, z) = E_0\pi b \int_{-p_0}^{p_0} u'(z + \pi b p)\underset{\eta_1}{\overline{\tau}_{rzj}}(\mu, \gamma, \delta, \lambda, p)\,dp$$

where $u_{av} = 0.5[u(z - 1) + u(z + 1)]$, or as a Stieltjes integral

$$\underset{T}{\sigma_{ij}}(\mu, \gamma, \delta, r, z) = E_0\left[u_{av} \cdot \underset{1}{\overline{\sigma}_{ij}}(\mu, \gamma, \delta, \lambda) - \int_{-p_0}^{p_0} \underset{\eta_1}{\overline{\sigma}_{ij}}(\mu, \gamma, \delta, \lambda, p)\,du(z + \pi b p)\right]$$

$$\underset{T}{\tau_{rzj}}(\mu, \gamma, \delta, r, z) = E_0 \int_{-p_0}^{p_0} \underset{\eta_1}{\overline{\tau}_{rzj}}(\mu, \gamma, \delta, \lambda, p)\,du(z + \pi b p)$$

Numerical Methods for Calculating the Thermal Stresses in Composite Cylinders with Continuous or Discontinuous Temperature Functions T(z)

1. Recommendations for the Practical Use of the Tables of Stresses for the Temperature Function $\eta_1(z)$. The Accuracy Needed in These Tables.

In this chapter we use our approximate solutions in the development of practical methods for calculating thermal stresses for any variation of the temperature $T(z)$ along the cylinder. We use the thermal stresses $\underset{\eta_1}{\sigma_{ij}}$ and $\underset{\eta_1}{\tau_{rzj}}$ ($i = r, \theta, z; j = 0, 1$) in the shell and the core of a composite cylinder that are given in the appended tables for the temperature function $\eta_1(z)$ [cf. (0.4)] and expressed as multiples of $a_0 E_0$. These tables are for composite cylinders with the following parameters: $\mu = a/b = 0.7(0.05)1$, $\gamma = a_1/a_0 = 0.5(0.5)2$, and $\delta = E_1/E_0 = 0.5, 1$, and 2, where a and b are the internal and external radii of the shell, a_0 and a_1 are the thermal-expansion coefficients, and E_0 and E_1 are the moduli of elasticity of the materials making up the shell and the core respectively. The parameter values were so selected that the results would be applicable to composite cylinders that are used or could be used in transportation structures. The stresses $\underset{\eta_1}{\sigma_{ij}}$ and $\underset{\eta_1}{\tau_{rzj}}$ are tabulated for various values of the independent variables $\lambda = r/b$ and $p = z/\pi b$, so that they can be used in the calculation of thermal stresses for temperature functions $T(z)$ in cylinders of various dimensions. The variable λ plays the role of a parameter, and varies from 0 to μ in steps of 0.1 and from μ to 1 in steps of 0.05. For convenience in setting up the tables, the reference point for the temperature function $\eta_1(z)$ [cf. (0.4)] was taken to be the midpoint of the composite cylinder, and the independent variable $p = z/\pi b$ was measured from 0 to $H/2\pi b$, where H is the length of the cylinder. The normal stresses σ_{ij} ($i = r, \theta, z; j = 0, 1$) are odd and the tangential stresses $\underset{\eta_1}{\tau_{rzj}}$ ($j = 0, 1$) even functions of p, and for negative values $-p$ from 0 to $-H/2\pi b$, for all values of the parameters μ, γ, and δ and of the independent variable λ, we have

$$\left.\begin{array}{l} \underset{\eta_1}{\sigma_{ij}}(\mu, \gamma, \delta, \lambda, -p) = -\underset{\eta_1}{\sigma_{ij}}(\mu, \gamma, \delta, \lambda, p) \\[2em] \underset{\eta_1}{\tau_{rzj}}(\mu, \gamma, \delta, \lambda, -p) = \underset{\eta_1}{\tau_{rzj}}(\mu, \gamma, \delta, \lambda, p) \end{array}\right\} \tag{5.1}$$

The stresses $\underset{\eta_1}{\sigma_{ij}}$ and $\underset{\eta_1}{\tau_{rzj}}$ as multiples of $a_0 E_0$ are given in the tables with an absolute error $|\Delta|_{\eta_1} \leq 0.0005$ for values of the independent variable p from 0 to $p_0 = 0.8$. For p from 0.8 to $H/2\pi b$, these stresses, with an error $|\Delta|_{\eta_1}$, are independent of p, i.e., constant for any values of μ, γ, δ, and λ and equal to the values of the stresses for $p = p_0 = 0.8$.

As is known [see (2.36)], for $p \geq p_0$ the stresses are given, with an error $|\Delta|_{\eta_1}$, by the formulas

$$\sigma_{ij}(\mu, \gamma, \delta, \lambda, p \geq p_0) = \tfrac{1}{2}\sigma_{ij}(\mu, \gamma, \delta, \lambda)$$
$$\underset{\eta_1}{} \qquad\qquad\qquad\qquad\qquad \underset{1}{}$$

$$\tau_{rzj}(\mu, \gamma, \delta, \lambda, p \geq p_0) = \tfrac{1}{2}\tau_{rzj}(\mu, \gamma, \delta, \lambda) = 0$$
$$\underset{\eta_1}{} \qquad\qquad\qquad\qquad\qquad \underset{1}{}$$

$$(i = r, \theta, z; \ j = 0, 1) \tag{5.2}$$

where σ_{ij} and τ_{rzj} are the stresses in a composite cylinder for the temperature function $T = 1$, ob-
$\quad\underset{1}{}\qquad\underset{1}{}$
tained from (2.46) and (2.47).

It is clear that the larger $|\Delta|_{\eta_1}$, the smaller will be the value of p_0 for which (5.2) will be satisfied.

Hence the value of p_0 and, consequently, the length of the cylinder H depend on the absolute error $|\Delta|_{\eta_1}$ in the calculated stresses σ_{ij} and τ_{rzj}. The results in the tables are for $|\Delta|_{\eta_1} \leq 0.0005$, for
$\qquad\qquad\qquad\qquad\qquad\qquad\quad \underset{\eta_1}{} \quad\quad \underset{\eta_1}{}$
which $p_0 = 0.8$. In this case the length H must not be less than $2p_0\pi b$, i.e., $H \geq 2p_0\pi b = 1.6\pi b$. If the error is larger than 0.0005 in practical calculations, then from the tables of σ_{ij} and τ_{rzj} for specific
$\qquad\qquad\qquad\qquad\qquad\qquad\qquad\qquad\qquad\qquad\qquad\qquad\qquad\qquad\quad \underset{\eta_1}{} \qquad \underset{\eta_1}{}$
values of μ, γ, and δ we can easily find the corresponding value of p_0 and the lower bound for the cylinder length. In a specific cylinder, for a given error $|\Delta|_{\eta_1}$, the value of p_0 varies with varying $\lambda = r/b$, as can readily be seen from the tables. Thus, with our method, stresses can be calculated more accurately for long cylinders than for short cylinders.

We will now show how to calculate in practice the absolute error $|\Delta|_{\eta_1}$ of the calculated values of the stresses σ_{ij} and τ_{rzj} and a lower bound for the length of a composite cylinder for which we can
$\qquad\quad \underset{\eta_1}{} \qquad \underset{\eta_1}{}$
calculate the stresses with an absolute error less than $|\Delta|_T$ for a temperature function $T(z)$. From (3.14) it follows that the maximum absolute error in the approximate formulas (3.11) or (3.12) for stresses with $T = T(z)$ is

$$|\Delta|_T = a_0 E_0 |\Delta|_{\eta_1} \cdot \bigvee_{-H/2}^{H/2} (T) \tag{5.3}$$

where $\displaystyle\bigvee_{-H/2}^{H/2} (T)$ is the total variation of the temperature function $T(z)$ in the interval $[-H/2, H/2]$, that is, along the whole length of the cylinder.

For a specific cylinder with known parameters μ, γ, and δ and known temperature distribution $T(z)$, the quantity $a_0 E_0 \displaystyle\bigvee_{-H/2}^{H/2} (T)$ is easily calculated. For a specified maximum permissible error $|\Delta|_T$, we find $|\Delta|_{\eta_1}$ from (5.3) and use the tables of stresses for this cylinder to determine p_0 and the lower bound for the cylinder length $H \geq 2p_0\pi b$. It is natural that the error $|\Delta|_T$ can only be such that $|\Delta|_{\eta_1}$ is not less than 0.0005.

The Calculation of the Total Variation of the Temperature
Along the Whole Length of the Cylinder

We use one end of the cylinder as the initial point for reckoning the temperature, and write the total variation $\displaystyle\bigvee_{0}^{H} (T)$. In practice this quantity is calculated as follows. The interval $[0, H]$ is divided by

45

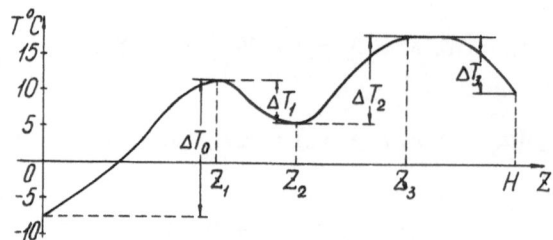

Fig. 12. The calculation of the total variation of
the temperature $T(z)$.

the points z_i into n subintervals (in the general case, of arbitrary length) in each of which $T(z)$ is monotonic, i.e., either nondecreasing or nonincreasing (see Fig. 12).

The sum of the absolute values of the variations of $T(z)$ in the subintervals is $\overset{H}{\underset{0}{V}}(T)$, and so

$$\overset{H}{\underset{0}{V}}(T) = \sum_{i=0}^{n-1} |T(z_{i+1}) - T(z_i)| = \sum_{i=0}^{n-1} |\Delta T_i|$$

Here $z_0 = 0$ and $z_n = H$.

For the temperature function shown in Fig. 12, $n = 4$ and $\overset{H}{\underset{0}{V}}(T) = |\Delta T_0| + |\Delta T_1| + |\Delta T_2|$
$+ |\Delta T_3| = 19 + 6 + 13 + 7 = 45°$. If $T(z)$ is monotonic everywhere in $[0, H]$, i.e., if it is either nondecreasing or nonincreasing along the whole length of the cylinder, then

$$\overset{H}{\underset{0}{V}}(T) = |T(H) = T(0)|$$

In (5.3) $|\Delta|_T$ is the maximum error of all the approximate formulas (3.11) or (3.12) in the calculation of stresses for $T = T(z)$. Calculations using these formulas can either be carried out exactly, in which case the integrals are taken in finite form, or approximately, with any desired accuracy. In the choice of $|\Delta|_T$ we must take into account the possible error in the calculation of the integrals in (3.11) or (3.12). Below, in the description of methods for calculating stresses given by the formulas (3.11) or (3.12), we will dwell briefly on the errors in these methods. If we denote the maximum absolute error made in the calculation of the integrals in (3.11) or (3.12) by $|\delta|_T$, the total maximum absolute error in stresses calculated from these formulas will be

$$|\Delta| = |\Delta|_T + |\delta|_T . \tag{5.4}$$

Example 3. We are required to find the absolute error $|\Delta|_{\eta_1}$ permissible in the calculated stresses σ_{ij} and τ_{rzj} ($i = r, \theta, z$; $j = 0, 1$) if the stresses in a composite cylinder with the temperature function $\eta_1 \qquad \eta_1$ shown in Fig. 12 must be determined with an error less than $|\Delta| = 1$ kg/cm². For concrete, such an error is 5% of the ultimate tensile strength $\sigma_{ul} = 20$ kg/cm². For the shell material we may assume $a_0 = 10 \cdot 10^{-6}$ deg⁻¹ and $E_0 = 200{,}000$ kg/cm². The total variation of the temperature function in

46

Fig. 12 is $\overset{H}{\underset{0}{\bigvee}}(T) = 45°$, and $a_0 E_0 \overset{H}{\underset{0}{\bigvee}}(T) = 90 \text{ kg/cm}^2$. In (5.4) we set $|\Delta|_T = 0.5 \text{ kg/cm}^2$ and $|\delta|_T = 0.5 \text{ kg/cm}^2$. From (5.3) we obtain

$$|\Delta|_{\eta_1} = \frac{|\Delta|_T}{a_0 E_0 \overset{H}{\underset{0}{\bigvee}}(T)} = \frac{0.5}{90} > 0.005$$

We note in passing that $|\Delta|_{\eta_1}$ is dimensionless. Hence the stresses $\underset{\eta_1}{\sigma_{ij}}$ and $\underset{\eta_1}{\tau_{rzj}}$ are obtained from the appended tables with a maximum error of 0.005. The method used for calculating the integrals (3.11) or (3.12) must yield results with errors not greater than $|\delta|_T = 0.5 \text{ kg/cm}^2$.

After the error $|\Delta|_{\eta_1}$ has been found, it is recommended that graphs of the stresses $\underset{\eta_1}{\sigma_{ij}}$ and $\underset{\eta_1}{\tau_{rzj}}$ ($i = r, \theta, z; j = 0, 1$) for a definite composite cylinder be plotted. In drawing the graphs, we can use either z or $p = z/\pi b$ as the independent variable and employ the formulas

$$\underset{\eta_1}{\sigma_{ij}}(\mu, \gamma, \delta, r, z) = \underset{\eta_1}{\sigma_{ij}}(\mu, \gamma, \delta, \lambda, p)$$

$$\underset{\eta_1}{\tau_{rzj}}(\mu, \gamma, \delta, r, z) = \underset{\eta_1}{\tau_{rzj}}(\mu, \gamma, \delta, \lambda, p)$$

$$(i = r, \theta, z; j = 0, 1)$$

As an example we use a composite cylinder with $\mu = 0.9$, $\gamma = 0.5$, and $\delta = 0.5$, and obtain the stresses $\underset{\eta_1}{\sigma_{ij}}$ and $\underset{\eta_1}{\tau_{rzj}}$ for a temperature function $\eta_1(z)$ on the cylinder axis ($\lambda = 0$) and on the cylinder surface ($\lambda = 1$), and the stress $\underset{\eta_1}{\tau_{rzj}}$ for $\lambda = 0.75$. For other values of the variable $\lambda = r/b$ the graphs are obtained in similar manner. We use $p = z/\pi b$ as the independent variable and obtain values directly from the tables for $\mu = 0.9$, $\gamma = 0.5$, and $\delta = 0.5$ with an error $|\Delta|_{\eta_1} \leq 0.005$ (see Fig. 13).

The properties of the stresses $\underset{\eta_1}{\sigma_{ij}}$ and $\underset{\eta_1}{\tau_{rzj}}$ obtained from (5.1) and (5.2) are clearly shown in the graphs. With an error $|\Delta|_{\eta_1} \leq 0.005$, the stresses become constant for $p \geq p_0 = 0.6$. If z is used as the independent variable in the graphs, it is convenient to express it in units of πb:

$$z = p\pi b \tag{5.5}$$

In this case we find $p = z/\pi b$ for each z and look up the corresponding stresses $\underset{\eta_1}{\sigma_{ij}}$ and $\underset{\eta_1}{\tau_{rzj}}$ in the tables. It is more convenient to give the value of p and then to use (5.5) to find z. If each abscissa $0p$ in Fig. 13 is multiplied by πb, the $0p$ axis becomes the $0z$ axis and we obtain the graph with the independent variable z. In plotting the graphs of $\underset{\eta_1}{\sigma_{ij}}$ and $\underset{\eta_1}{\tau_{rzj}}$ in Fig. 13, we used the midpoint of the cylinder for the origin of p. In general, the origin of p or z can be taken at any point of the cylinder, and the calculated stresses are independent of this choice. For example, if, for the same temperature function (a unit step function with the discontinuity at the middle of the cylinder), the origin of the independent variable p is taken at the point $p = -0.5$ in the original coordinate system [the new T and $(1/a_0 E_0)\underset{\eta_1}{\sigma_{ij}}$ axes are shown by dotted lines], then the temperature and stress graphs remain unchanged.

In practical calculations, it is convenient to use the left end of the cylinder as the zero point of z or p, and we shall do so in the following sections.

47

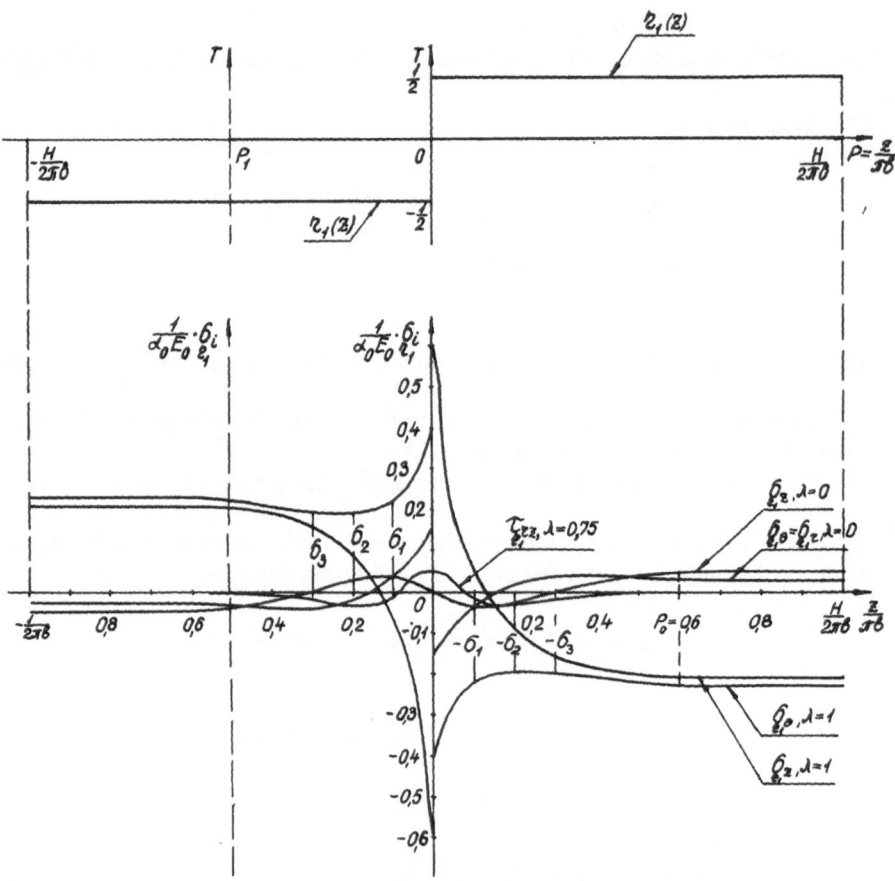

Fig. 13. Thermal stresses in a composite cylinder with the parameters $\mu = 0.9$, $\gamma = 0.5$, and $\delta = 0.5$, for the temperature function $\eta_1(z)$.

2. Calculation of Thermal Stresses in a Composite Cylinder with an Arbitrary Temperature Discontinuity

We will show how to use the thermal stresses σ_{ij} and τ_{rzj} ($i = r, \theta, z; j = 0, 1$) in a composite cylinder with a unit temperature discontinuity $\eta_1(z)$ for the calculation of the stresses in the case of an arbitrary temperature discontinuity.

Let the temperature function be an arbitrary step function (see Fig. 14, Curve 1):

$$T(z) = \begin{cases} T_1 & 0 \leq z < z_0 \\ \\ T_2 & z_0 < z \leq H \end{cases} \tag{5.6}$$

where T_1 and T_2 are arbitrary constants and z_0 satisfies (2.34).

The temperature function (5.6) can be written in the form

$$T(z) = \tfrac{1}{2}(T_1 + T_2) + (T_2 - T_1)\eta_1(z - z_0) \tag{5.7}$$

that is. as the sum of the average temperature $T_{av} = \tfrac{1}{2}(T_1 + T_2)$ (see Fig. 14, Curve 2) and the step function $\Delta T = T_2 - T_1$ (see Fig. 14, Curve 3).

48

It is plain that the thermal stresses can also be expressed as sums:

$$\underset{T}{\sigma_{ij}}(\mu, \gamma, \delta, r, z) = T_{av} \cdot \underset{1}{\sigma_{ij}}(\mu, \gamma, \delta, r) + \Delta T \underset{\eta_1}{\sigma_{ij}}(\mu, \gamma, \delta, r, z - z_0)$$

$$\underset{T}{\tau_{rzj}}(\mu, \gamma, \delta, r, z) = \Delta T \cdot \underset{\eta_1}{\tau_{rzj}}(\mu, \gamma, \delta, r, z - z_0)$$

$$(i = r, \theta, z; \ j = 0, 1)$$

(5.8)

where the $\underset{1}{\sigma_{ij}}$ are the stresses for $T = 1$ given by (2.46) and (2.47) and $\underset{\eta_1}{\sigma_{ij}}(\mu, \gamma, \delta, r, z - z_0)$
$= \underset{\eta_1}{\sigma_{ij}}(\mu, \gamma, \delta, \lambda, p - z_0/\pi b)$ and $\underset{\eta_1}{\tau_{rzj}}(\mu, \gamma, \delta, r, z - z_0) = \underset{\eta_1}{\tau_{rzj}}(\mu, \gamma, \delta, \lambda, p - z_0/\pi b)$ are the stresses
for the temperature function $\eta_1(z)$ translated to the right by an amount z_0.

In Fig. 14, as an example, we show only the tangential stresses $\underset{T}{\sigma_{\theta 0}}$ at the surface of a composite
cylinder ($\lambda = 1$) with parameters $\mu = 0.9$, $\gamma = 0.5$, and $\delta = 0.5$ obtained from (5.8) (see Fig. 13). The
first term $T_{av}\underset{1}{\sigma_{\theta 0}}$ in (5.8) is shown by Curve 4 (Fig. 14), the second term $\Delta T \underset{\eta_1}{\sigma_{\theta 0}}(\mu, \gamma, \delta, r, z - z_0)$ by
Curve 5. Curve 5 was obtained from the graph of $\underset{\eta_1}{\sigma_{\theta 0}}$ for $\lambda = 1$ in Fig. 13 by multiplying each of the
ordinates $\sigma_1, \sigma_2, \sigma_3, \ldots, -\sigma_1, -\sigma_2, -\sigma_3, \ldots$ by ΔT and translating the axis a distance z_0 to the
right. We note that the temperature discontinuity given by (5.6) for $\Delta T = T_2 - T_1 > 0$ is shown in
Fig. 14. We will say that this type of temperature discontinuity is positive.

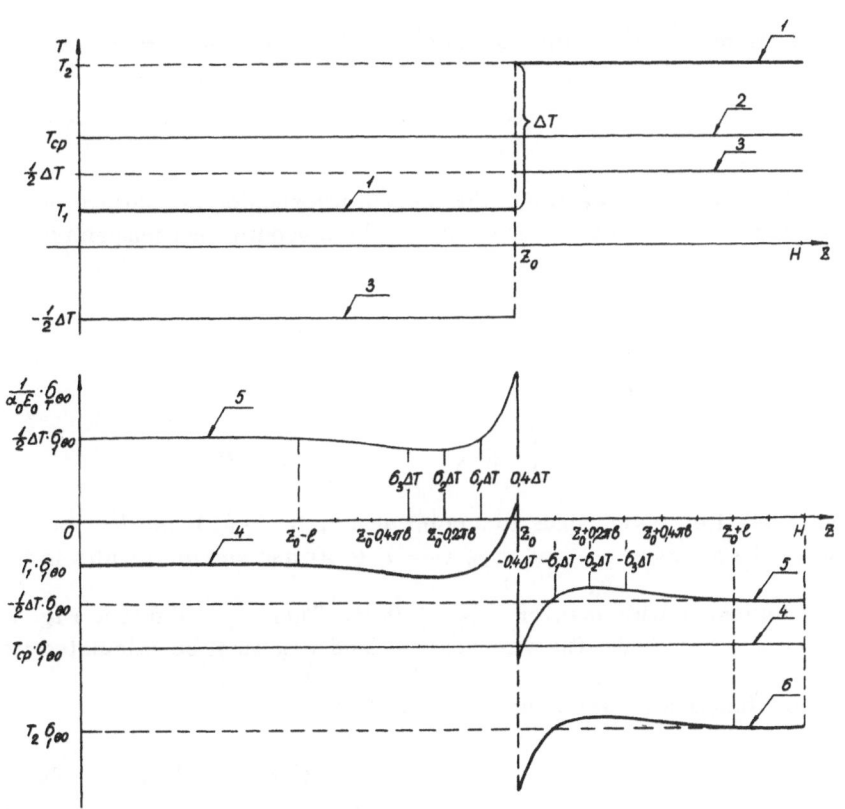

Fig. 14. Tangential stresses on the surface of a composite cylinder ($\mu = 0.9$,
$\gamma = 0.5$, $\delta = 0.5$) for an arbitrary temperature discontinuity $\Delta T = T_2 - T_1$;
$l = p_0\pi b = 0 = 0.6\pi b$; $T_{av} = \frac{1}{2}(T_1 + T_2)$.

If the property (5.2) of the stresses $\underset{\eta_1}{\sigma_{ij}}$ and $\underset{\eta_1}{\tau_{rzj}}$ is used, the stresses (5.8) can be written

$$\underset{T}{\sigma_{ij}}(\mu, \gamma, \delta, r, z) = \begin{cases} \underset{1}{T_1\sigma_{ij}}(\mu, \gamma, \delta, r) & 0 \leq z < z_0 - l,\ l = p_0\pi b \\[2ex] \underset{1}{T_{av}\sigma_{ij}}(\mu, \gamma, \delta, r) + \underset{\eta_1}{\Delta T\sigma_{ij}}(\mu, \gamma, \delta, r, z - z_0) \\[2ex] \hspace{6cm} z_0 - l \leq z \leq z_0 + l \\[2ex] \underset{1}{T_2\sigma_{ij}}(\mu, \gamma, \delta, r) & z_0 + l \leq z \leq H \end{cases}$$

(5.9)

$$\underset{T}{\tau_{rzj}}(\mu, \gamma, \delta, r, z) = \begin{cases} \underset{\eta_1}{\Delta T \cdot \tau_{rzj}}(\mu, \gamma, \delta, r, z - z_0) & z_0 - l \leq z \leq z_0 + l \\[2ex] 0 & 0 \leq z \leq z_0 - l,\ z_0 + l \leq z \leq H \end{cases}$$

The tangential stress $\underset{T}{\sigma_{\theta 0}}$ (5.9) on the surface of a composite cylinder ($\lambda = 1$) is shown by Curve 6 in Fig. 14; the ordinates of this curve are the sums of the ordinates of Curves 4 and 5. The absolute error of stresses calculated from (5.9) does not exceed the value

$$|\Delta| = |\Delta|_T = |T_2 - T_1| \cdot |\Delta|_{\eta_1} \cdot a_0 E_0$$

(5.10)

calculated from (5.4). Here $|\delta|_T = 0$, since the stresses (5.9) are expressed exactly in terms of $\underset{\eta_1}{\sigma_{ij}}$ and $\underset{\eta_1}{\tau_{rzj}}$.

Example 4. In this example we calculate the thermal stresses on the surface of a composite cylinder with parameters $\mu = 0.9$ and $\gamma = \delta = 0.5$ for the discontinuous temperature function (Fig. 15, Curve 1)

$$T(z) = \begin{cases} 5°C & 0 \leq z < z_0 \\[2ex] -20°C & z_0 < z \leq H \end{cases}$$

where z_0 satisfies (2.34), i.e., corresponds to a point at a distance at least $l = p_0\pi b$ from the ends.

We will calculate the stresses $\underset{\eta_1}{\sigma_{i0}}$ and $\underset{\eta_1}{\tau_{rz0}}$ ($i = r, \theta, z$) and use the results in the formulas (5.9) to obtain the required stresses with an error $|\Delta|_{\eta_1} \leq 0.01$. Then $p_0 = 0.6$ (see Fig. 13). On the surface of the cylinder $\underset{T}{\sigma_{r0}} = \underset{T}{\tau_{rz0}} = 0$. The stresses $\underset{T}{\sigma_{\theta 0}}$ and $\underset{T}{\sigma_{z0}}$ must be calculated from (5.8) and (5.9). The temperature function we are using can be written

$$T(z) = -7.5 - 25 \cdot \eta_1(z) = -7.5 - 25 \cdot \begin{cases} -\frac{1}{2} & 0 \leq z < z_0 \\[2ex] \frac{1}{2} & z_0 < z \leq H \end{cases}$$

50

i.e., as the sum of the mean temperature $T_{av} = \frac{1}{2}(T_1 + T_2) = \frac{1}{2}(5 - 20) = -7.5°C$ (see Fig. 15, Curve 2) and the negative discontinuity $\Delta T = T_2 - T_1 = -20 - 5 = -25°C$ (see Fig. 15 Curve 3). In other words, we set $T_1 = 5°C$ and $T_2 = -20°C$ in (5.6) and (5.7).

Substituting $T_{av} = -7.5°C$ and $\Delta T = -25°C$ in (5.8), we obtain the required stresses

$$\underset{T}{\sigma_{\theta 0}} = -7.5 \cdot \underset{1}{\sigma_{\theta 0}} - 25 \cdot \underset{\eta_1}{\sigma_{\theta 0}(z - z_0)}$$

$$\underset{T}{\sigma_{z 0}} = -7.5 \cdot \underset{1}{\sigma_{z 0}} - 25 \cdot \underset{\eta_1}{\sigma_{z 0}(z - z_0)}$$

From (2.47), with the given parameters and $r = b$, we obtain $\underset{1}{\sigma_{\theta 0}} = -0.46 \, a_0 E_0$ and $\underset{1}{\sigma_{z 0}} = -0.42 \, a_0 E_0$.

These same values, divided by 2, are also given in the tables of the stresses $\underset{\eta_1}{\sigma_{\theta 0}}$ and $\underset{\eta_1}{\sigma_{z 0}}$ for $p \geq 0.6$ (see Fig. 13).

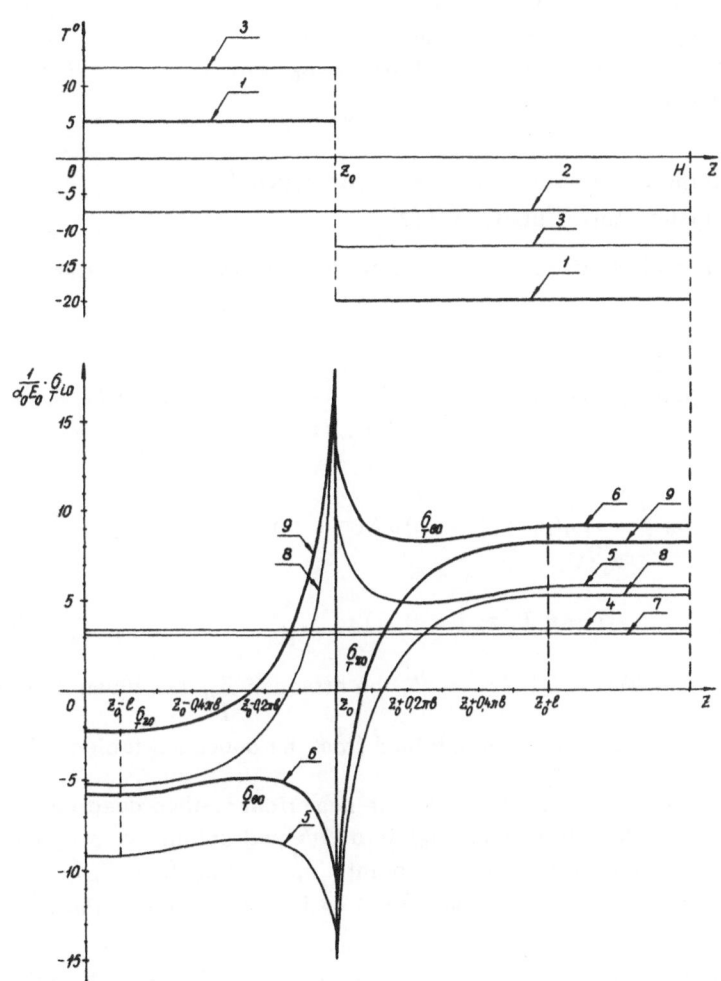

Fig. 15. Thermal stresses on the surface of a composite cylinder ($\mu = 0.9$, $\gamma = 0.5$, $\delta = 0.5$) for the negative temperature discontinuity $\Delta T = -25°$.

For the constant mean temperature, we obtain

$$-7.5 \cdot \underset{1}{\sigma_{\theta 0}} = -7.5 \cdot (-0.46)\, a_0 E_0 = 3.45\, a_0 E_0 \qquad \text{(see Fig. 15, Curve 4)}$$

$$-7.5 \cdot \underset{1}{\sigma_{z0}} = -7.5 \cdot (-0.42)\, a_0 E_0 = 3.15\, a_0 E_0 \qquad \text{(see Fig. 15, Curve 7)}$$

The stresses $\underset{\eta_1}{\sigma_{\theta 0}}(z)$ and $\underset{\eta_1}{\sigma_{z0}}(z)$ for $\lambda = 1$ are obtained from the appended tables for the parameters $\mu = 0.9$ and $\gamma = \delta = 0.5$. These stresses are plotted in Fig. 13. The multiplication of the ordinates of the curves for the stresses $\underset{\eta_1}{\sigma_{\theta 0}}(z)$ and $\underset{\eta_1}{\sigma_{z0}}(z)$ by -25 yields the functions $-25\underset{\eta_1}{\sigma_{\theta 0}}(z)$ and $-25\underset{\eta_1}{\sigma_{z0}}(z)$. We also plot the graphs of these functions translated a distance z_0 to the right, i.e., graphs with the origin at the point z_0. The graphs of the functions $-25\underset{\eta_1}{\sigma_{\theta 0}}(z - z_0)$ (see Fig. 15, Curve 5) and $-25\underset{\eta_1}{\sigma_{z0}}(z - z_0)$ (see Fig. 15, Curve 8) are obtained similarly. Adding the ordinates of Curves 4 and 5, and those of Curves 7 and 8, we obtain Curves 6 and 9 for the stresses $\underset{T}{\sigma_{\theta 0}}$ and $\underset{T}{\sigma_{z0}}$. The analytical expressions for these stresses are given by (5.9) with $T_1 = 5°\text{C}$, $T_2 = -20°\text{C}$, $T_{av} = -7.5°\text{C}$, $\Delta T = -25°\text{C}$, $\mu = 0.9$, $\gamma = \delta = 0.5$, $\lambda = 1$, and $p_0 = 0.6$.

From (5.10), the errors do not exceed $25 \cdot 0.01 \cdot a_0 E_0$, which is $1.25\, a_0 E_0 \%$ of the ultimate tensile strength $\sigma_{ul} = 20\ \text{kg/cm}^2$.

3. Calculation of Thermal Stresses in a Composite Cylinder for an Arbitrary Temperature Function T(z)

From (3.12), the thermal stresses in a composite cylinder for a temperature function $T(z)$ are given by Stieltjes integrals:

$$
\left.
\begin{aligned}
\underset{T}{\sigma_{ij}}(\mu, \gamma, \delta, r, z) &= a_0 E_0 \left[T_{av} \cdot \underset{1}{\overline{\sigma}_{ij}}(\mu, \gamma, \delta, \lambda) - \int_{-p_0}^{p_0} \underset{\eta_1}{\overline{\sigma}_{ij}}(\mu, \gamma, \delta, \lambda, p)\, dT(z + \pi b p) \right] \\[2ex]
\underset{T}{\tau_{rzj}}(\mu, \gamma, \delta, r, z) &= a_0 E_0 \int_{-p_0}^{p_0} \underset{\eta_1}{\overline{\tau}_{rzj}}(\mu, \gamma, \delta, \lambda, p)\, dT(z + \pi b p) \\[2ex]
& \qquad (i = r, \theta, z;\ j = 0, 1)
\end{aligned}
\right\} \qquad (5.11)
$$

where $T_{av} = \tfrac{1}{2}[T(z - l) + T(z + l)]$, $l = p_0 \pi b$, the stresses $\underset{1}{\overline{\sigma}_{ij}}$ are given as multiples of $a_0 E_0$ by (2.46) and (2.47), and the stresses $\underset{\eta_1}{\overline{\sigma}_{ij}}$ are obtained from the appended tables. To calculate the Stieltjes integrals (5.11), it is convenient to use the numerical method described below.

We divide the range of integration $[-p_0, p_0]$ into $2n$ equal subintervals by the points $\overline{p}_k = kh_0$ ($k = 0, 1, 2, \ldots, n$) to the right of zero and the points $\overline{p}_k' = -kh_0$ ($k = 0, 1, 2, \ldots, n$) to the left of zero, where $h_0 = p_0/n$ is the integration step. We note that $\overline{p}_n = nh_0 = p_0$, $\overline{p}_n' = -nh_0 = -p_0$, $\overline{p}_0 = \overline{p}_0' = 0$.

In the Stieltjes integrals (5.11), the point z at which we are to calculate the stresses is a parameter, since the integration is with respect to the independent variable p. We fix z in the integrals in (5.11), setting $z = z^*$, where z^* is an arbitrary fixed point at a distance at least $p_0 \pi b$ from the ends of the interval $[0, H]$.

When the independent variable $p = z/\pi b$ in the integrals in (5.11) varies over $[-p_0, p_0]$, the argument z in the temperature function $T(z)$ varies over the interval $[z^* - l, z^* + l]$. When $[-p_0, p_0]$ is divided into $2n$ subintervals, the interval $[z^* - l, z^* + l]$ is also divided into $2n$ parts by the points $z_k = z^* + kh_0\pi b = z^* + kh$ $(k = 0, 1, 2, \ldots, n)$ to the right of z^* and the points $z'_k = z^* - kh_0\pi b = z^* - kh$ $(k = 0, 1, 2, \ldots, n)$ to the left of z^*, where $h = \pi bh_0 = \pi bp_0/n = l/n$. We note that $z_0 = z'_0 = z^*$, $z_n = z^* + l$, and $z'_n = z^* - l$. For given parameters μ, γ, δ, and λ, we consider the values of the stresses $\overline{\sigma}_{ij}(\mu, \gamma, \delta, \lambda, p)$ and $\overline{\tau}_{rzj}(\mu, \gamma, \delta, \lambda, p)$ at the midpoints $\xi_k = \overline{p}_k + \tfrac{1}{2}h_0$
η_1 η_1
$= (k + \tfrac{1}{2})h_0$ of the intervals $[\overline{p}_k, \overline{p}_{k+1}]$ and at the midpoints $\xi'_k = \overline{p}'_{k+1} + \tfrac{1}{2}h_0 = -(k + 1)h_0 + \tfrac{1}{2}h_0$ $= -(k + \tfrac{1}{2})h_0$ of the intervals $[\overline{p}'_{k+1}, \overline{p}'_k]$. For the above division of $[-p_0, p_0]$ and the choice of midpoints ξ_k and ξ'_k of the subintervals obtained by this division, the Stieltjes integrals in (5.11) are given approximately by integral sums, and

$$
\begin{aligned}
\underset{T}{\sigma_{ij}}(\mu, \gamma, \delta, r, z^*) = a_0E_0\Bigg\{ &T_{av} \cdot \underset{1}{\overline{\sigma}_{ij}}(\mu, \gamma, \delta, \lambda) \\
&- \sum_{k=0}^{n-1}\{T(z^* - kh) - T[z^* - (k + 1)h]\}\underset{\eta_1}{\sigma_{ij}}[\mu, \gamma, \delta, \lambda, -(k + \tfrac{1}{2})h_0] \\
&- \sum_{k=0}^{n-1}\{T[z^* + (k + 1)h] - T(z^* + kh)\} \cdot \underset{\eta_1}{\sigma_{ij}}[\mu, \gamma, \delta, \lambda, (k + \tfrac{1}{2})h_0]\Bigg\} \\[6pt]
\underset{T}{\tau_{rzj}}(\mu, \gamma, \delta, r, z^*) = a_0E_0\Bigg\{ &\sum_{k=0}^{n-1}\{T(z^* - kh) - T[z^* - (k + 1)h]\}\underset{\eta_1}{\tau_{rzj}}[\mu, \gamma, \delta, \lambda, -(k + \tfrac{1}{2})h_0] \\
&+ \sum_{k=0}^{n-1}\{T[z^* + (k + 1)h] - T(z^* + kh)\}\underset{\eta_1}{\tau_{rzj}}[\mu, \gamma, \delta, \lambda, (k + \tfrac{1}{2})h_0]\Bigg\} \\
&(i = r, \theta, z; j = 0, 1)
\end{aligned}
\tag{5.12}
$$

If in (5.12) we replace z^* by z, use the notation $T'_k = T(z - kh)$, $T'_{k+1} = T[z - (k + 1)h]$, $\Delta T'_k = T'_k - T'_{k+1}$, $T_k = T(z + kh)$, $T_{k+1} = T[z + (k + 1)h]$, and $\Delta T_k = T_{k+1} - T_k$, the relations (5.2), and the identities $-\underset{\eta_1}{\sigma_{ij}}[\mu, \gamma, \delta, \lambda, -(k + \tfrac{1}{2})h_0] = \underset{\eta_1}{\sigma_{ij}}[\mu, \gamma, \delta, \lambda, (k + \tfrac{1}{2})h_0]$ and $\underset{\eta_1}{\tau_{rzj}}[\mu, \gamma, \delta, \lambda, -(k + \tfrac{1}{2})h_0]$
$= \underset{\eta_1}{\tau_{rzj}}[\mu, \gamma, \delta, \lambda, (k + \tfrac{1}{2})h_0]$, we obtain

$$
\begin{aligned}
\underset{T}{\sigma_{ij}}(\mu, \gamma, \delta, r, z) &= a_0E_0\Bigg\{T_{av} \cdot \underset{1}{\overline{\sigma}_{ij}}(\mu, \gamma, \delta, \lambda) + \sum_{k=0}^{n-1}(\Delta T'_k - \Delta T_k) \cdot \underset{\eta_1}{\overline{\sigma}_{ij}}[\mu, \gamma, \delta, \lambda, (k + \tfrac{1}{2})h_0]\Bigg\} \\[6pt]
\underset{T}{\tau_{rzj}}(\mu, \gamma, \delta, r, z) &= a_0E_0\sum_{k=0}^{n-1}(\Delta T'_k + \Delta T_k)\underset{\eta_1}{\overline{\tau}_{rzj}}[\mu, \gamma, \delta, (k + \tfrac{1}{2})h_0]
\end{aligned}
\tag{5.13}
$$

where $T_{av} = \tfrac{1}{2}[T(z - l) + T(z + l)]$, $l \le z \le H - l$, and $l = p_0\pi b$.

We now give an example and use it to interpret geometrically the method of calculating the stresses at an arbitrary point $z = z_0$ from the formulas (5.13).

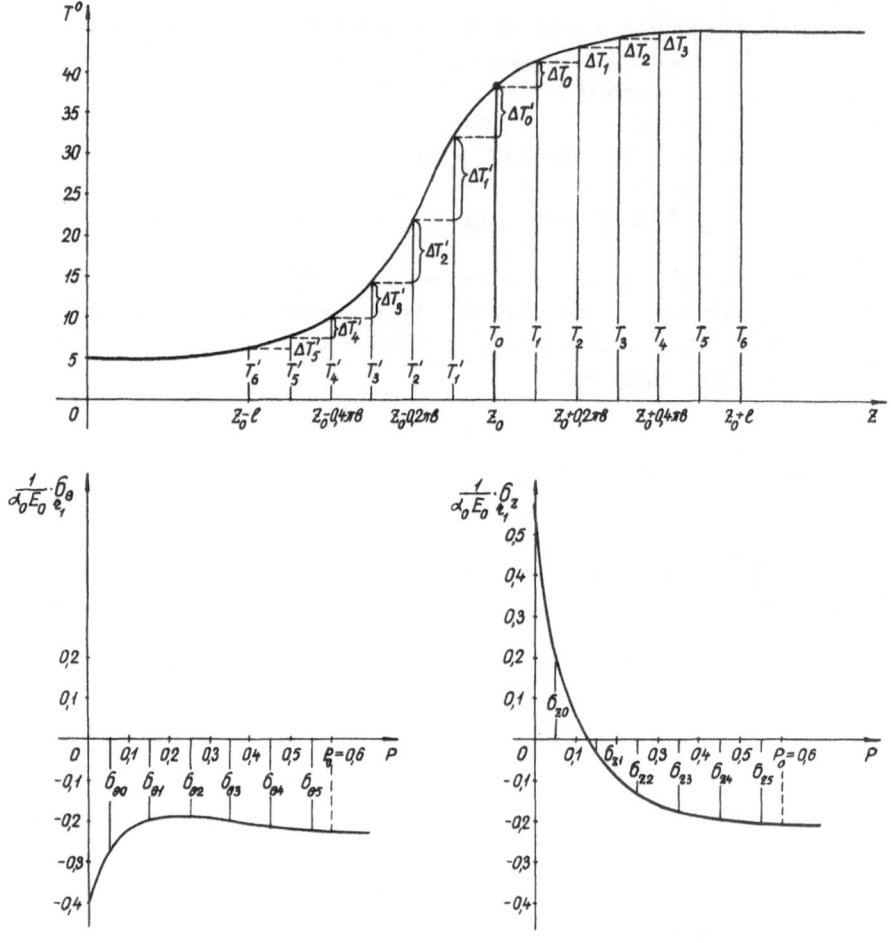

Fig. 16. The calculation of thermal stresses at an arbitrary point z_0
for a continuous temperature function $T(z)$.

Example 5. We find the thermal stresses on the surface of a composite cylinder with parameters $\mu = 0.9$ and $\gamma = \delta = 0.5$ at an arbitrary point $z = z_0$ for the temperature function shown in Fig. 16. On the surface we have $\sigma_{r0} = \tau_{rz0} = 0$, and we need only calculate $\sigma_{\theta 0}$ and σ_{z0}. Using the tables, we plot the stresses $\overline{\sigma}_{\theta 0}$ and $\overline{\sigma}_{z0}$ for $\lambda = 1$, $\mu = 0.9$, and $\gamma = \delta = 0.5$ with an error $|\Delta|_{\eta_1} < 0.005$ (see Fig. 16). To this accuracy, the stresses $\overline{\sigma}_{\theta 0}$ and $\overline{\sigma}_{z0}$ remain constant for $p_0 = 0.6$. We divide the interval $[0, p_0]$ into $n = 6$ subintervals of length $h_0 = p_0/n = 0.6/6 = 0.1$. Thus, the interval $[z_0 - l, z_0 + l]$, with $l = p_0\pi b = 0.6\pi b$, is divided into $2n = 12$ subintervals of length $h = l/n = 0.6\pi b/6 = 0.1\pi b$. From the graph $T(z)$ we find T_k at the points $z_k = z_0 + 0.1\pi bk$ and T_k' at the points $z_k' = z_0 - 0.1\pi bk$ $(k = 0, 1, 2, \ldots, 6)$. We also find the increments in the ordinates $\Delta T_k = T_{k+1} - T_k$ and $\Delta T_k' = T_k' - T_{k+1}'$ $(k = 0, 1, 2, \ldots, 5)$. We now use the graphs of $\sigma_{\theta 0}$ and σ_{z0} to find the ordinates $\sigma_{\theta k}$ and σ_{zk} at the points $p_k = 0.1(k + \frac{1}{2})$ $(k = 0, 1, 2, \ldots, 5)$ (see Fig. 16). The stresses $\overline{\sigma}_{\theta 0}$ and $\overline{\sigma}_{z0}$ are obtained from (2.47) for $r = b$, and are $\overline{\sigma}_{\theta 0} = -0.46$ and $\overline{\sigma}_{z0} = -0.42$.

From (5.13)

$$\sigma_{\theta 0} \atop T = a_0 E_0 \left[0.5(T_6' + T_6)\overline{\sigma}_{\theta 0} \Big|_1 + \sum_{k=0}^{5}(\Delta T_k' - \Delta T_k)\sigma_{\theta k} \right] = a_0 E_0 [-0.5(6 + 45) \cdot 0.46 - (6 - 3)0.27$$

$$- (10 - 2) \cdot 0.2 - (8 - 1)0.19 - (4.5 - 0.5)0.2 - (2.5 - 0)0.22 - (1.5 - 0)0.23] = -16.97\, a_0 E_0$$

$$\sigma_{z 0} \atop T = a_0 E_0 \left[0.5(T_6' + T_6)\overline{\sigma}_{z 0} \Big|_1 + \sum_{k=0}^{5}(\Delta T_k' - \Delta T_k)\sigma_{z k} \right] = a_0 E_0 [-0.5(6 + 45)0.42 + (6 - 3)0.2$$

$$- (10 - 2) \cdot 0.02 - (8 - 1)0.13 - (4.5 - 0.5) \cdot 0.8 - (2.5 - 0)0.2 - (1.5 - 0)0.21] = -14.40\, a_0 E_0$$

The Error in the Numerical Calculation of the
Stieltjes Integrals (5.11)

The absolute error $|\delta|_T$ in the calculation of the integrals in (5.11) by the approximate formulas (5.13) becomes smaller when the integration step $h_0 = p_0/n$ decreases. The number of divisions n of the interval $[0, p_0]$ can always be selected so that $|\delta|_T$ will be small enough. In practice, $|\delta|_T$ can be determined by a second calculation. The stresses (5.11) are calculated with steps h_0 and $h_0/2$, and the results are compared. If these results differ by less than $|\delta|_T$, we can assume that the actual error is less than the same quantity; if they do not agree with this precision, then we decrease the step by one-half, repeat the calculation with the step $h_0/4$, and again compare the results obtained with steps $h_0/2$ and $h_0/4$, etc.

Repeated calculations have the further advantage that they eliminate errors of calculation, since these errors are easily detected by comparing the results obtained for the steps h_0 and $h_0/2$.

4. Calculation of Thermal Stresses in a Composite Cylinder for a Bounded Discontinuous Temperature Function T(z)

Let the temperature distribution in a composite cylinder be given by a temperature function $T(z)$ having a finite discontinuity for $z = z_0$ (Fig. 17). We write $T(z_0 - 0) = T_1$, $T(z_0 + 0) = T_2$, and

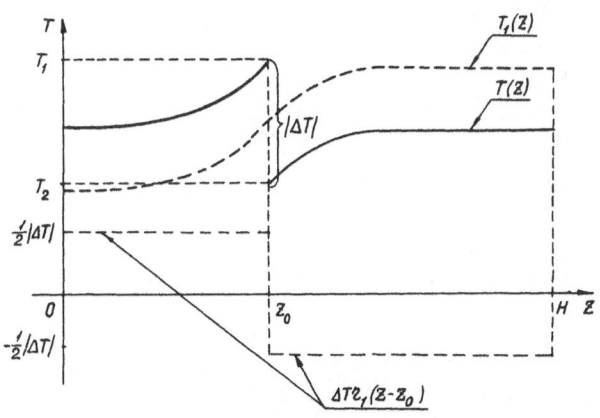

Fig. 17. A discontinuous temperature function $T(z)$ expressed as the sum of a continuous function $T_1(z)$ and a discontinuous function $\Delta T \eta(z - z_0)$.

55

$\Delta T = T_2 - T_1$. We express $T(z)$ as the sum of two functions

$$T(z) = T_1(z) + \Delta T \cdot \eta_1(z - z_0) \qquad (5.14)$$

where $T_1(z)$ is the continuous function obtained by vertically displacing the parts of the function $T(z)$ to the left and to the right of the discontinuity so that the points (z_0, T_1) and (z_0, T_2) are displaced to the same point $\left(z_0, \frac{1}{2}(T_1 + T_2)\right)$. In other words, the function $T_1(z)$ is obtained from $T(z)$ by adding to every ordinate of the latter to the left of z_0 the constant $\frac{1}{2}\Delta T$ and to every ordinate to the right of z_0 the constant $-\frac{1}{2}\Delta T$ (see Fig. 17). The quantity $\Delta T \eta_1(z - z_0)$ is the discontinuity at z_0 of $T = T_2 - T_1$ (see Fig. 17), where $\eta_1(z - z_0)$ is the unit step function:

$$\eta_1(z - z_0) = \begin{cases} -\frac{1}{2} & 0 \leq z < z_0 \\[2ex] \frac{1}{2} & z_0 < z \leq H \end{cases}$$

The quantity ΔT can be either positive or negative. In Fig. 17 we have $\Delta T < 0$, and this type of discontinuity is called negative.

The calculation of the thermal stresses in a composite cylinder for a continuous temperature function $T_1(z)$ was described in Section 3 of this chapter, and that for an arbitrary discontinuity $\Delta T \eta_1(z - z_0)$ in Section 2. The thermal stresses for the discontinuous temperature function (5.14) are calculated by combining the stresses obtained for the functions $T_1(z)$ and $\Delta T \eta_1(z - z_0)$.

If a temperature function has not only one discontinuity, as in Fig. 17, but n finite discontinuities, then it can be expressed as above as the sum of one continuous function and n step functions.

Conclusion

Our investigation of thermal stresses in shells filled with concrete, or, in our terminology, composite cylinders, yields methods for the practical calculation of thermal stresses in cylindrical structures. These methods can be used in planning, and in the development of methods for preventing thermal cracks in cylindrical structures. This can be done by the appropriate choice of the material used for the shell and core and by the correct technological preparation of these materials. Work on problems such as increasing the life of structures used in transportation installations has been carried out by TsNIIS, and in particular in the hydraulic- and electric-analog laboratory of Professor V. S. Luk'yanov.

The author wishes to thank A. A. Abramov for his active participation in the seminars at VTs AN SSSR, where this work was developed, and for his suggestions concerning the mathematical part of the work which were incorporated in its final published form.

The author also would like to thank Professor V. S. Luk'yanov, I. I. Denisov, M. D. Golovko, R. I. Veitsman, and L. N. Losev for reading the manuscript and making useful suggestions for its improvement, and in particular Professor Luk'yanov for his assistance in the writing of Chapter 5.

Literature Cited

1. Silin, K. S., Glotov, N. M., Gretsov, A. P., Karpinskii, V. I., and Prokhorov, A. D. *Bridge Support Foundations Constructed of Prefabricated Concrete Shells.* Moscow, Transzheldorizdat, 1958.
2. Nikishin, V. S., and Losev, L. N. "Thermal Stresses in Reinforced-Concrete Shells Filled with Concrete." *Transportnoe Stroetel'stvo,* 1961, No. 7.
3. Luk'yanov, V. S., and Denisov, I. I. *The Prevention of Thermal Cracks in Concrete Bridge Supports.* Moscow, Transzheldorizdat, 1959.
4. Vvedenskaya, N. D., and Shnol', É. É. "A Method for Calculating Stresses in Circular Cylinders." In the collection *Numerical Mathematics,* No. 7, Moscow, Izd. AN SSSR, 1961.
5. Danilovskaya, V. I. "Application of the Castigliano Variational Theorem to Three-Dimensional Thermoelastic Problems." *Archiwum mechaniki stosowanej,* 1962, 5, 14.
6. Nikishin, V. S. *The Effect on Thermally-Stressed Solid or Hollow Cylinders of Arbitrary Temperature Distributions along Their Length.* Moscow, VTs AN SSSR, 1962.
7. Melan, É., and Parkus, G. *Thermal Stresses Generated by Stationary Temperature Fields.* Moscow, Fizmatgiz, 1958.
8. Gatewood, B. E. *Thermal Stresses* [Russian translation]. Moscow, Izd. In. Lit., 1959.
9. Natanson, I. P. *The Theory of Functions of a Real Variable.* Moscow, Gostekhizdat, 1950.

Appendix

Thermal Stresses in the Shell and Core of a Composite Cylinder

$$\overline{\sigma}_{ij}(\mu, \gamma, \delta, r, z) = \frac{1}{a_0 E_0} \, \sigma_{ij}(\mu, \gamma, \delta, r, z)$$
$$\eta_1 \qquad\qquad\qquad\qquad \eta_1$$

$$\overline{\tau}_{rzj}(\mu, \gamma, \delta, r, z) = \frac{1}{a_0 E_0} \, \tau_{rzj}(\mu, \gamma, \delta, r, z)$$
$$\eta_1 \qquad\qquad\qquad\qquad \eta_1$$

$$(i = r, \theta, z; \, j = 0, 1)$$

for the temperature function

$$\eta_1(z) = \begin{cases} \tfrac{1}{2} & 0 < z \le \dfrac{H}{2} \\[2ex] -\tfrac{1}{2} & -\dfrac{H}{2} \le z < 0 \end{cases}$$

with the parameter values $\mu = a/b = 0.7(0.05)0.95$, $\gamma = a_1/a_0 = 0.5(0.5)2$, $\delta = E_1/E_0 = 0.5, 1$, and 2, $\lambda = r/b = 0, 0.3, 0.6$, and $0.7(0.05)0.95$, and the independent variable $p = z/\pi b = 0, 0.01, 0.02, 0.05(0.05)0.3$, and $0.3(0.1)0.8$. The Poisson ratio used was $\nu = 1/6$ for concrete.

Stresses were calculated with an absolute-value error $|\Delta|_{\eta_1} \le 0.0005$.

Exact values are given for the normal stresses $\overline{\sigma}_{ij}$ $(i = r, \theta, z; \, j = 0, 1)$ on the section $z = 0$ of
$$\eta_1$$
the cylinder.

	λ → P ↓	Core				Shell						
		0	0.3	0.6	0.7	0.7	0.75	0.80	0.85	0.90	0.95	1.00
$\dfrac{\bar{\sigma}_z}{\eta_1}$	0	0	0	0	0.116	−0.004	0	0	0	0	0	0.6
	0.01	−0.003	−0.002	0.013	0.096	−0.055	−0.080	−0.060	−0.050	−0.042	−0.006	0.562
	0.02	−0.006	−0.004	0.021	0.079	−0.106	−0.121	−0.104	−0.084	−0.052	0.067	0.516
	0.05	−0.013	−0.009	0.019	0.040	−0.206	−0.174	−0.139	−0.083	0.013	0.173	0.387
	0.10	−0.020	−0.013	0.018	0.042	−0.223	−0.166	−0.105	−0.034	0.049	0.143	0.238
	0.15	−0.016	−0.006	0.036	0.066	−0.180	−0.133	−0.083	−0.030	0.025	0.081	0.134
	0.20	−0.002	0.010	0.058	0.086	−0.141	−0.110	−0.076	−0.042	−0.007	0.025	0.055
	0.25	0.018	0.031	0.075	0.098	−0.117	−0.098	−0.077	−0.056	−0.036	−0.018	−0.002
	0.30	0.038	0.050	0.087	0.104	−0.105	−0.093	−0.081	−0.069	−0.058	−0.049	−0.042
	0.40	0.073	0.080	0.100	0.108	−0.097	−0.094	−0.090	−0.087	−0.085	−0.084	−0.084
	0.50	0.092	0.096	0.105	0.108	−0.098	−0.097	−0.096	−0.097	−0.097	−0.098	−0.099
	0.60	0.101	0.103	0.106	0.107	−0.099	−0.099	−0.100	−0.100	−0.101	−0.102	−0.103
	0.70	0.105	0.105	0.106	0.106	−0.101	−0.101	−0.101	−0.101	−0.102	−0.102	−0.103
	0.80	0.106	0.106	0.106	0.106	−0.102	−0.102	−0.102	−0.102	−0.102	−0.102	−0.102
$\dfrac{\bar{\sigma}_\theta}{\eta_1}$	0	−0.15	−0.15	−0.15	−0.159	−0.554	−0.6	−0.6	−0.6	−0.6	−0.6	−0.4
	0.01	−0.136	−0.135	−0.135	−0.146	−0.542	−0.550	−0.543	−0.532	−0.512	−0.467	−0.319
	0.02	−0.122	−0.121	−0.119	−0.127	−0.514	−0.505	−0.489	−0.468	−0.434	−0.375	−0.272
	0.05	−0.081	−0.078	−0.070	−0.065	−0.418	−0.386	−0.355	−0.321	−0.281	−0.236	−0.189
	0.10	−0.020	−0.016	0.001	0.012	−0.283	−0.251	−0.222	−0.194	−0.168	−0.143	−0.120
	0.15	0.025	0.030	0.045	0.053	−0.206	−0.181	−0.158	−0.139	−0.121	−0.106	−0.091
	0.20	0.055	0.058	0.068	0.072	−0.169	−0.149	−0.132	−0.117	−0.105	−0.094	−0.084
	0.25	0.070	0.072	0.077	0.079	−0.156	−0.139	−0.124	−0.113	−0.103	−0.094	−0.087
	0.30	0.077	0.078	0.079	0.079	−0.155	−0.139	−0.126	−0.116	−0.107	−0.099	−0.093
	0.40	0.076	0.076	0.075	0.074	−0.165	−0.150	−0.137	−0.127	−0.119	−0.112	−0.107
	0.50	0.071	0.071	0.070	0.069	−0.175	−0.160	−0.147	−0.137	−0.128	−0.121	−0.115
	0.60	0.067	0.067	0.066	0.066	−0.181	−0.165	−0.153	−0.142	−0.133	−0.126	−0.120
	0.70	0.065	0.065	0.064	0.064	−0.184	−0.168	−0.155	−0.145	−0.136	−0.128	−0.121
	0.80	0.064	0.064	0.064	0.064	−0.186	−0.170	−0.156	−0.146	−0.136	−0.129	−0.122
$\dfrac{\bar{\sigma}_r}{\eta_1}$	0	−0.15	−0.15	−0.15	−0.319	−0.319	−0.6	−0.6	−0.6	−0.6	−0.6	0
	0.01	−0.136	−0.135	−0.144	−0.253	−0.253	−0.389	−0.428	−0.414	−0.356	−0.202	0
	0.02	−0.122	−0.120	−0.132	−0.196	−0.196	−0.258	−0.284	−0.259	−0.182	−0.055	0
	0.05	−0.081	−0.076	−0.069	−0.062	−0.062	−0.067	−0.056	−0.032	−0.005	0.008	0
	0.10	−0.020	−0.012	0.018	0.036	0.036	0.028	0.025	0.022	0.016	0.008	0
	0.15	0.025	0.033	0.054	0.062	0.062	0.047	0.035	0.024	0.015	0.006	0
	0.20	0.055	0.059	0.067	0.067	0.067	0.051	0.037	0.024	0.014	0.006	0
	0.25	0.070	0.071	0.070	0.067	0.067	0.051	0.036	0.024	0.014	0.006	0
	0.30	0.077	0.076	0.070	0.067	0.067	0.050	0.035	0.023	0.013	0.006	0
	0.40	0.076	0.074	0.067	0.065	0.065	0.048	0.035	0.023	0.014	0.006	0
	0.50	0.071	0.069	0.065	0.064	0.064	0.048	0.034	0.023	0.014	0.006	0
	0.60	0.067	0.066	0.064	0.064	0.064	0.048	0.034	0.023	0.014	0.007	0
	0.70	0.065	0.065	0.064	0.064	0.064	0.047	0.034	0.023	0.014	0.007	0
	0.80	0.064	0.064	0.064	0.064	0.064	0.047	0.034	0.023	0.014	0.007	0
$\dfrac{\bar{\tau}_{rs}}{\eta_1}$	0	0	0.012	−0.014	$-\infty$	$-\infty$	−0.011	0.106	0.189	0.258	0.322	0
	0.01	0	0.012	−0.009	−0.048	−0.048	0.020	0.107	0.176	0.224	0.211	0
	0.02	0	0.011	−0.001	0.004	0.004	0.055	0.103	0.143	0.151	0.083	0
	0.05	0	0.009	0.014	0.026	0.026	0.045	0.046	0.031	0.003	−0.022	0
	0.10	0	0.000	−0.009	−0.016	−0.016	−0.025	−0.034	−0.041	−0.041	−0.029	0
	0.15	0	−0.010	−0.030	−0.040	−0.040	−0.049	−0.052	−0.050	−0.041	−0.025	0
	0.20	0	−0.018	−0.038	−0.044	−0.044	−0.048	−0.048	−0.043	−0.034	−0.019	0
	0.25	0	−0.020	−0.037	−0.038	−0.038	−0.040	−0.038	−0.033	−0.025	−0.014	0
	0.30	0	−0.019	−0.030	−0.030	−0.030	−0.030	−0.027	−0.023	−0.017	−0.009	0
	0.40	0	−0.012	−0.017	−0.016	−0.016	−0.014	−0.013	−0.010	−0.007	−0.004	0
	0.50	0	−0.006	−0.007	−0.006	−0.006	−0.006	−0.005	−0.004	−0.002	−0.001	0
	0.60	0	−0.002	−0.003	−0.002	−0.002	−0.002	−0.001	−0.001	−0.001	0.000	0
	0.70	0	−0.001	−0.001	0.000	0.000	0.000	0.000	0.000	0.000	0.000	0
	0.80	0	0.000	0.000	0.000	0.000	0.000	0.000	0.000	0.000	0.000	0

$\gamma = 0,5; \mu = 0,75; \delta = 0,5$

	λ	Core					Shell					
	p	0	0.3	0.6	0.7	0.75	0.75	0.80	0.85	0.90	0.95	1.00
$\bar{\sigma}_z$ η_1^z	0	0	0	0	0	0.116	−0.004	0	0	0	0	0.6
	0.01	−0.003	−0.003	0.004	0.031	0.094	−0.065	−0.085	−0.063	−0.051	−0.014	0.552
	0.02	−0.007	−0.006	0.007	0.038	0.073	−0.122	−0.128	−0.104	−0.067	0.053	0.496
	0.05	−0.016	−0.014	0.006	0.026	0.038	−0.217	−0.166	−0.107	−0.010	0.147	0.355
	0.10	−0.025	−0.019	0.009	0.033	0.051	−0.205	−0.138	−0.064	0.020	0.113	0.207
	0.15	−0.022	−0.013	0.028	0.057	0.076	−0.159	−0.110	−0.058	−0.003	0.051	0.103
	0.20	−0.008	0.003	0.049	0.075	0.091	−0.130	−0.099	−0.066	−0.034	−0.003	0.025
	0.25	0.010	0.023	0.065	0.086	0.098	−0.116	−0.097	−0.078	−0.060	−0.043	−0.029
	0.30	0.030	0.042	0.076	0.092	0.100	−0.111	−0.100	−0.090	−0.080	−0.072	−0.066
	0.40	0.063	0.070	0.089	0.096	0.099	−0.112	−0.110	−0.107	−0.105	−0.104	−0.104
	0.50	0.082	0.085	0.093	0.096	0.097	−0.116	−0.116	−0.116	−0.116	−0.117	−0.118
	0.60	0.090	0.092	0.095	0.095	0.096	−0.119	−0.120	−0.120	−0.121	−0.121	−0.122
	0.70	0.094	0.094	0.095	0.095	0.095	−0.121	−0.121	−0.121	−0.122	−0.122	−0.123
	0.80	0.095	0.095	0.095	0.095	0.095	−0.122	−0.122	−0.122	−0.122	−0.122	−0.122
$\bar{\sigma}_\theta$ η_1	0	−0.15	−0.15	−0.15	−0.15	−0.159	−0.554	−0.6	−0.6	−0.06	−0.6	−0.4
	0.01	−0.136	−0.135	−0.133	−0.135	−0.143	−0.537	−0.544	−0.534	−0.515	−0.471	−0.322
	0.02	−0.123	−0.121	−0.117	−0.118	−0.121	−0.507	−0.494	−0.473	−0.440	−0.381	−0.279
	0.05	−0.082	−0.079	−0.068	−0.061	−0.057	−0.405	−0.370	−0.334	−0.294	−0.249	−0.202
	0.10	−0.023	−0.018	−0.001	0.010	0.016	−0.276	−0.244	−0.214	−0.187	−0.161	−0.138
	0.15	0.021	0.026	0.040	0.048	0.052	−0.208	−0.183	−0.162	−0.143	−0.127	−0.112
	0.20	0.048	0.052	0.061	0.065	0.067	−0.178	−0.159	−0.143	−0.129	−0.117	−0.106
	0.25	0.063	0.065	0.069	0.071	0.071	−0.169	−0.152	−0.139	−0.127	−0.118	−0.110
	0.30	0.069	0.070	0.071	0.071	0.071	−0.169	−0.154	−0.142	−0.132	−0.123	−0.116
	0.40	0.068	0.068	0.067	0.066	0.066	−0.180	−0.165	−0.154	−0.144	−0.136	−0.129
	0.50	0.063	0.063	0.062	0.061	0.061	−0.190	−0.175	−0.163	−0.153	−0.144	−0.137
	0.60	0.059	0.059	0.058	0.058	0.058	−0.195	−0.180	−0.168	−0.157	−0.149	−0.141
	0.70	0.057	0.057	0.057	0.057	0.057	−0.198	−0.183	−0.170	−0.159	−0.151	−0.143
	0.80	0.056	0.056	0.056	0.056	0.056	−0.199	−0.184	−0.171	−0.160	−0.151	−0.144
$\bar{\sigma}_r$ η_1	0	−0.15	−0.15	−0.15	−0.15	−0.319	−0.319	−0.6	−0.6	−0.6	−0.6	0
	0.01	−0.136	−0.135	−0.135	−0.160	−0.245	−0.245	−0.372	−0.396	−0.350	−0.200	0
	0.02	−0.122	−0.120	−0.118	−0.143	−0.180	−0.180	−0.229	−0.233	−0.171	−0.052	0
	0.05	−0.082	−0.076	−0.059	−0.050	−0.041	−0.041	−0.037	−0.019	0.002	0.010	0
	0.10	−0.023	−0.014	0.017	0.033	0.042	0.042	0.033	0.027	0.019	0.009	0
	0.15	0.021	0.029	0.049	0.055	0.058	0.058	0.043	0.029	0.017	0.007	0
	0.20	0.048	0.053	0.060	0.060	0.059	0.059	0.043	0.029	0.017	0.007	0
	0.25	0.063	0.064	0.062	0.060	0.058	0.058	0.042	0.028	0.016	0.007	0
	0.30	0.069	0.068	0.062	0.059	0.057	0.057	0.041	0.027	0.016	0.007	0
	0.40	0.068	0.066	0.060	0.057	0.056	0.056	0.040	0.027	0.016	0.007	0
	0.50	0.063	0.062	0.058	0.056	0.056	0.056	0.040	0.027	0.016	0.007	0
	0.60	0.059	0.059	0.057	0.056	0.056	0.056	0.040	0.027	0.017	0.008	0
	0.70	0.057	0.057	0.056	0.056	0.056	0.056	0.040	0.028	0.017	0.008	0
	0.80	0.056	0.056	0.056	0.056	0.056	0.056	0.040	0.028	0.017	0.008	0
$\bar{\tau}_{rz}$ η_1	0	0	0.016	0.012	−0.035	$-\infty$	$-\infty$	0.017	0.141	0.230	0.308	0
	0.01	0	0.016	0.013	−0.018	−0.027	−0.027	0.045	0.134	0.197	0.198	0
	0.02	0	0.015	0.015	0.005	0.018	0.018	0.071	0.113	0.130	0.072	0
	0.05	0	0.012	0.016	0.020	0.024	0.024	0.033	0.023	−0.003	−0.025	0
	0.10	0	0.001	−0.008	−0.018	−0.024	−0.024	−0.036	−0.043	−0.042	−0.029	0
	0.15	0	−0.010	−0.030	−0.039	−0.042	−0.042	−0.049	−0.048	−0.040	−0.025	0
	0.20	0	−0.017	−0.037	−0.042	−0.043	−0.043	−0.044	−0.040	−0.032	−0.018	0
	0.25	0	−0.019	−0.035	−0.036	−0.036	−0.036	−0.035	−0.030	−0.023	−0.013	0
	0.30	0	−0.018	−0.029	−0.029	−0.027	−0.027	−0.025	−0.022	−0.016	−0.009	0
	0.40	0	−0.011	−0.016	−0.015	−0.014	−0.014	−0.012	−0.010	−0.007	−0.004	0
	0.50	0	−0.005	−0.007	−0.006	−0.006	−0.006	−0.005	−0.004	−0.002	−0.001	0
	0.60	0	−0.002	−0.003	−0.002	−0.002	−0.002	−0.002	−0.001	−0.001	0.000	0
	0.70	0	−0.001	−0.001	−0.001	−0.001	−0.001	0.000	0.000	0.000	0.000	0
	0.80	0	0.000	0.000	0.000	0.000	0.000	0.000	0.000	0.000	0.000	0

	λ	Core				Shell						
	p	0	0.3	0.6	0.7	0.7	0.75	0.80	0.85	0.90	0.95	1.00
$\dfrac{\bar{\sigma}_z}{\eta_1}$	0	0	0	0	0.001	0.084	0	0	0	0	0	0.6
	0.01	−0.009	−0.009	−0.007	−0.021	0.028	−0.051	−0.044	−0.038	−0.030	0.006	0.577
	0.02	−0.018	−0.019	−0.018	−0.041	−0.026	−0.076	−0.074	−0.060	−0.028	0.092	0.546
	0.05	−0.044	−0.045	−0.057	−0.081	−0.132	−0.111	−0.084	−0.031	0.066	0.231	0.456
	0.10	−0.078	−0.080	−0.084	−0.079	−0.151	−0.093	−0.030	0.044	0.133	0.236	0.346
	0.15	−0.096	−0.094	−0.074	−0.052	−0.102	−0.052	0.003	0.062	0.126	0.191	0.257
	0.20	−0.098	−0.090	−0.054	−0.029	−0.058	−0.021	0.018	0.058	0.100	0.141	0.180
	0.25	−0.087	−0.076	−0.036	−0.014	−0.028	−0.004	0.022	0.048	0.073	0.097	0.119
	0.30	−0.071	−0.059	−0.023	−0.005	−0.011	0.004	0.020	0.036	0.051	0.064	0.075
	0.40	−0.038	−0.030	−0.008	0.001	0.003	0.007	0.012	0.017	0.021	0.024	0.026
	0.50	−0.017	−0.012	−0.002	0.002	0.004	0.005	0.006	0.007	0.007	0.007	0.006
	0.60	−0.006	−0.004	0.000	0.001	0.003	0.002	0.002	0.002	0.001	0.001	0.000
	0.70	−0.002	−0.001	0.000	0.000	0.001	0.001	0.001	0.000	0.000	0.000	0.000
	0.80	0.000	0.000	0.000	0.000	0.000	0.000	0.000	0.000	0.000	0.000	0.000
$\dfrac{\bar{\sigma}_\theta}{\eta_1}$	0	−0.3	−0.3	−0.3	−0.318	−0.554	−0.6	−0.6	−0.6	−0.6	−0.6	−0.4
	0.01	−0.282	−0.281	−0.281	−0.296	−0.525	−0.538	−0.533	−0.522	−0.502	−0.458	−0.310
	0.02	−0.264	−0.263	−0.261	−0.270	−0.484	−0.480	−0.468	−0.448	−0.416	−0.357	−0.254
	0.05	−0.212	−0.209	−0.198	−0.191	−0.354	−0.330	−0.304	−0.274	−0.237	−0.192	−0.145
	0.10	−0.134	−0.128	−0.106	−0.092	−0.176	−0.154	−0.132	−0.110	−0.087	−0.065	−0.044
	0.15	−0.073	−0.067	−0.046	−0.035	−0.068	−0.055	−0.042	−0.029	−0.017	−0.005	0.006
	0.20	−0.031	−0.026	−0.012	−0.005	−0.011	−0.004	0.003	0.009	0.015	0.021	0.026
	0.25	−0.006	−0.003	0.005	0.008	0.016	0.018	0.021	0.024	0.026	0.028	0.030
	0.30	0.006	0.008	0.011	0.013	0.025	0.025	0.026	0.027	0.027	0.028	0.028
	0.40	0.012	0.012	0.011	0.011	0.021	0.020	0.020	0.019	0.018	0.018	0.017
	0.50	0.008	0.008	0.007	0.006	0.012	0.011	0.011	0.010	0.009	0.009	0.008
	0.60	0.004	0.004	0.003	0.003	0.006	0.005	0.005	0.004	0.004	0.004	0.003
	0.70	0.002	0.002	0.001	0.001	0.002	0.002	0.002	0.001	0.001	0.001	0.001
	0.80	0.000	0.000	0.000	0.000	0.000	0.000	0.000	0.000	0.000	0.000	0.000
$\dfrac{\bar{\sigma}_r}{\eta_1}$	0	−0.3	−0.3	−0.3	−0.407	−0.407	−0.6	−0.6	−0.6	−0.6	−0.6	0
	0.01	−0.282	−0.281	−0.284	−0.342	−0.342	−0.418	−0.440	−0.421	−0.359	−0.203	0
	0.02	−0.264	−0.261	−0.264	−0.283	−0.283	−0.303	−0.307	−0.271	−0.188	−0.057	0
	0.05	−0.212	−0.205	−0.177	−0.145	−0.145	−0.124	−0.092	−0.053	−0.016	0.004	0
	0.10	−0.134	−0.121	−0.069	−0.038	−0.038	−0.026	−0.013	−0.002	0.003	0.003	0
	0.15	−0.073	−0.060	−0.022	−0.007	−0.007	−0.004	0.000	0.001	0.001	0.000	0
	0.20	−0.031	−0.022	−0.003	0.002	0.002	0.002	0.002	0.001	0.000	−0.001	0
	0.25	−0.006	−0.002	0.003	0.003	0.003	0.003	0.002	0.001	−0.001	−0.001	0
	0.30	0.006	0.007	0.005	0.003	0.003	0.002	0.001	0.000	−0.001	−0.001	0
	0.40	0.012	0.010	0.004	0.002	0.002	0.001	0.000	0.000	−0.001	−0.001	0
	0.50	0.008	0.006	0.002	0.001	0.001	0.000	0.000	0.000	0.000	0.000	0
	0.60	0.004	0.003	0.001	0.000	0.000	0.000	0.000	0.000	0.000	0.000	0
	0.70	0.002	0.001	0.000	0.000	0.000	0.000	0.000	0.000	0.000	0.000	0
	0.80	0.000	0.000	0.000	0.000	0.000	0.000	0.000	0.000	0.000	0.000	0
$\dfrac{\bar{\tau}_{rz}}{\eta_1}$	0	0	0.044	0.085	$-\infty$	$-\infty$	0.126	0.200	0.255	0.302	0.345	0
	0.01	0	0.044	0.088	0.121	0.121	0.148	0.198	0.241	0.267	0.233	0
	0.02	0	0.043	0.093	0.134	0.134	0.166	0.188	0.205	0.192	0.104	0
	0.05	0	0.039	0.088	0.105	0.105	0.117	0.107	0.079	0.036	−0.005	0
	0.10	0	0.025	0.036	0.025	0.025	0.012	−0.003	−0.017	−0.025	−0.021	0
	0.15	0	0.006	−0.006	−0.019	−0.019	−0.031	−0.038	−0.040	−0.035	−0.022	0
	0.20	0	−0.008	−0.026	−0.034	−0.034	−0.041	−0.043	−0.040	−0.033	−0.019	0
	0.25	0	−0.015	−0.031	−0.035	−0.035	−0.038	−0.038	−0.034	−0.026	−0.015	0
	0.30	0	−0.016	−0.029	−0.030	−0.030	−0.031	−0.029	−0.025	−0.019	−0.011	0
	0.40	0	−0.012	−0.018	−0.018	−0.018	−0.017	−0.015	−0.012	−0.009	−0.005	0
	0.50	0	−0.007	−0.009	−0.008	−0.008	−0.007	−0.006	−0.005	−0.003	−0.002	0
	0.60	0	−0.003	−0.004	−0.003	−0.003	−0.003	−0.002	−0.001	−0.001	0.000	0
	0.70	0	−0.001	−0.001	−0.001	−0.001	−0.001	0.000	0.000	0.000	0.000	0
	0.80	0	0.000	0.000	0.000	0.000	0.000	0.000	0.000	0.000	0.000	0

$\gamma = 1,0 \,; \mu = 0,95 \, \delta = 0,5$

		Core									Shell	
	λ	0	0.3	0.6	0.7	0.75	0.80	0.85	0.90	0.95	0.95	1.00
	p											
$\bar{\sigma}_z$ η_1	0.	0	0	0	0	0	0	0	0	0.001	0.084	0.6
	0.01	−0.010	−0.010	−0.012	−0.013	−0.013	−0.014	−0.015	−0.016	−0.041	−0.057	0.486
	0.02	−0.019	−0.020	−0.023	−0.024	−0.025	−0.026	−0.025	−0.023	−0.005	−0.004	0.465
	0.05	−0.046	−0.047	−0.049	−0.047	−0.043	−0.035	−0.018	0.017	0.085	0.168	0.412
	0.10	−0.079	−0.078	−0.064	−0.046	−0.030	−0.008	0.021	0.060	0.108	0.215	0.327
	0.15	−0.093	−0.088	−0.055	−0.029	−0.011	0.011	0.036	0.064	0.094	0.188	0.250
	0.20	−0.092	−0.082	−0.040	−0.015	0.000	0.017	0.035	0.054	0.074	0.147	0.183
	0.25	−0.080	−0.068	−0.028	−0.007	0.005	0.017	0.030	0.042	0.054	0.109	0.129
	0.30	−0.064	−0.053	−0.018	−0.002	0.006	0.015	0.023	0.031	0.038	0.077	0.087
	0.40	−0.035	−0.027	−0.007	0.001	0.005	0.008	0.012	0.014	0.017	0.034	0.036
	0.50	−0.015	−0.011	−0.002	0.001	0.003	0.004	0.005	0.006	0.006	0.012	0.012
	0.60	−0.005	−0.004	0.000	0.001	0.001	0.001	0.002	0.002	0.002	0.003	0.003
	0.70	−0.001	−0.001	0.000	0.000	0.000	0.000	0.000	0.000	0.000	0.000	0.000
	0.80	0.000	0.000	0.000	0.000	0.000	0.000	0.000	0.000	0.000	0.000	0.000
$\bar{\sigma}_\theta$ η_1	0	−0.3	−0.3	−0.3	−0.3	−0.3	−0.3	−0.3	−0.3	−0.318	−0.554	−0.4
	0.01	−0.281	−0.280	−0.276	−0.273	−0.270	−0.267	−0.263	−0.256	−0.244	−0.462	−0.326
	0.02	−0.262	−0.260	−0.252	−0.245	−0.241	−0.235	−0.227	−0.214	−0.190	−0.372	−0.270
	0.05	−0.207	−0.202	−0.184	−0.171	−0.163	−0.152	−0.139	−0.122	−0.100	−0.201	−0.150
	0.10	−0.127	−0.120	−0.096	−0.082	−0.074	−0.065	−0.055	−0.044	−0.033	−0.067	−0.045
	0.15	−0.067	−0.061	−0.041	−0.031	−0.026	−0.020	−0.014	−0.009	−0.003	−0.007	0.004
	0.20	−0.027	−0.023	−0.010	−0.004	−0.002	0.001	0.004	0.007	0.010	0.019	0.024
	0.25	−0.005	−0.002	0.005	0.008	0.009	0.010	0.012	0.013	0.014	0.028	0.029
	0.30	0.007	0.008	0.011	0.012	0.012	0.013	0.013	0.013	0.014	0.027	0.028
	0.40	0.011	0.011	0.011	0.010	0.010	0.010	0.010	0.009	0.009	0.019	0.018
	0.50	0.008	0.007	0.007	0.006	0.006	0.008	0.005	0.005	0.005	0.010	0.009
	0.60	0.004	0.004	0.003	0.003	0.003	0.004	0.002	0.002	0.002	0.004	0.004
	0.70	0.002	0.002	0.001	0.001	0.001	0.002	0.001	0.001	0.001	0.002	0.001
	0.80	0.000	0.000	0.000	0.000	0.000	0.000	0.000	0.000	0.000	0.000	0.000
$\bar{\sigma}_r$ η_1	0	−0.3	−0.3	−0.3	−0.3	−0.3	−0.3	−0.3	−0.3	−0.407	−0.407	0
	0.01	−0.281	−0.279	−0.268	−0.259	−0.252	−0.242	−0.227	−0.202	−0.136	−0.136	0
	0.02	−0.262	−0.257	−0.237	−0.220	−0.207	−0.189	−0.162	−0.118	−0.039	−0.039	0
	0.05	−0.207	−0.197	−0.153	−0.123	−0.102	−0.077	−0.048	−0.019	0.003	0.003	0
	0.10	−0.127	−0.112	−0.064	−0.040	−0.027	−0.016	−0.007	0.000	0.002	0.002	0
	0.15	−0.067	−0.055	−0.022	−0.011	−0.006	−0.003	−0.001	0.000	0.000	0.000	0
	0.20	−0.027	−0.020	−0.005	−0.001	0.000	0.000	0.000	0.000	−0.001	−0.001	0
	0.25	−0.005	−0.002	0.002	0.002	0.001	0.001	0.000	0.001	−0.001	−0.001	0
	0.30	0.007	0.007	0.005	0.003	0.002	0.001	0.000	0.000	−0.001	−0.001	0
	0.40	0.011	0.009	0.004	0.002	0.002	0.001	0.000	0.000	−0.001	−0.001	0
	0.50	0.008	0.006	0.003	0.001	0.001	0.000	0.000	0.000	0.000	0.000	0
	0.60	0.004	0.003	0.001	0.001	0.000	0.000	0.000	0.000	0.000	0.000	0
	0.70	0.002	0.001	0.001	0.000	0.000	0.000	0.000	0.000	0.000	0.000	0
	0.80	0.000	0.000	0.000	0.000	0.000	0.000	0.000	0.000	0.000	0.000	0
$\bar{\tau}_{rz}$ η_1	0	0	0.047	0.101	0.123	0.135	0.149	0.163	0.178	−∞	−∞	0
	0.01	0	0.047	0.100	0.121	0.132	0.144	0.156	0.169	0.181	0.181	0
	0.02	0	0.046	0.097	0.115	0.124	0.132	0.138	0.136	0.100	0.100	0
	0.05	0	0.039	0.074	0.080	0.079	0.073	0.060	0.038	0.011	0.011	0
	0.10	0	0.021	0.026	0.017	0.010	0.002	−0.006	−0.013	−0.014	−0.014	0
	0.15	0	0.003	−0.008	−0.016	−0.020	−0.023	−0.024	−0.023	−0.018	−0.018	0
	0.20	0	−0.009	−0.024	−0.028	−0.029	−0.029	−0.027	−0.023	−0.017	−0.017	0
	0.25	0	−0.014	−0.027	−0.029	−0.028	−0.026	−0.024	−0.019	−0.014	−0.014	0
	0.30	0	−0.015	−0.025	−0.025	−0.024	−0.022	−0.019	−0.015	−0.011	−0.011	0
	0.40	0	−0.010	−0.015	−0.014	−0.013	−0.012	−0.010	−0.008	−0.005	−0.005	0
	0.50	0	−0.006	−0.008	−0.007	−0.006	−0.006	−0.005	−0.004	−0.002	−0.002	0
	0.60	0	−0.003	−0.003	−0.003	−0.002	−0.002	−0.002	−0.001	−0.001	−0.001	0
	0.70	0	−0.001	−0.001	−0.001	−0.001	−0.001	−0.001	−0.001	0.000	0.000	0
	0.80	0	0.000	0.000	0.000	0.000	0.000	0.000	0.000	0.000	0.000	0

	λ / p	Core				Shell						
		0	0.3	0.6	0.7	0.7	0.75	0.80	0.85	0.90	0.95	1.00
$\dfrac{\bar\sigma_z}{\eta_1}$	0	0	0	0	−0.113	0.172	0	0	0	0	0	0.6
	0.01	−0.015	−0.017	−0.027	−0.138	0.110	−0.023	−0.028	−0.025	−0.019	0.019	0.593
	0.02	−0.031	−0.033	−0.056	−0.160	0.054	−0.030	−0.044	−0.035	−0.005	0.117	0.576
	0.05	−0.075	−0.081	−0.134	−0.202	−0.059	−0.048	−0.029	0.020	0.119	0.289	0.524
	0.10	−0.137	−0.146	−0.186	−0.200	−0.078	−0.020	0.044	0.122	0.218	0.330	0.453
	0.15	−0.177	−0.181	−0.184	−0.170	−0.025	0.030	0.089	0.155	0.226	0.301	0.379
	0.20	−0.193	−0.190	−0.166	−0.143	0.026	0.067	0.111	0.158	0.207	0.256	0.304
	0.25	−0.191	−0.183	−0.148	−0.125	0.062	0.090	0.120	0.151	0.182	0.213	0.241
	0.30	−0.180	−0.169	−0.134	−0.115	0.083	0.102	0.121	0.140	0.160	0.177	0.193
	0.40	−0.149	−0.140	−0.116	−0.105	0.102	0.108	0.115	0.121	0.127	0.132	0.135
	0.50	−0.126	−0.121	−0.108	−0.104	0.105	0.107	0.109	0.110	0.111	0.111	0.111
	0.60	−0.113	−0.111	−0.106	−0.104	0.105	0.104	0.104	0.104	0.104	0.103	0.102
	0.70	−0.108	−0.107	−0.106	−0.105	0.103	0.103	0.103	0.102	0.102	0.101	0.102
	0.80	−0.106	−0.106	−0.106	−0.106	0.101	0.101	0.101	0.101	0.101	0.101	0.101
$\dfrac{\bar\sigma_\theta}{\eta_1}$	0	−0.45	−0.45	−0.45	−0.476	−0.554	−0.6	−0.6	−0.6	−0.6	−0.6	−0.4
	0.01	−0.428	−0.428	−0.427	−0.447	−0.508	−0.525	−0.522	−0.512	−0.493	−0.449	−0.301
	0.02	−0.407	−0.405	−0.404	−0.414	−0.453	−0.456	−0.447	−0.428	−0.397	−0.339	−0.235
	0.05	−0.344	−0.340	−0.327	−0.318	−0.290	−0.274	−0.254	−0.227	−0.192	−0.149	−0.102
	0.10	−0.248	−0.241	−0.213	−0.196	−0.069	−0.057	−0.042	−0.026	−0.007	0.012	0.032
	0.15	−0.171	−0.163	−0.137	−0.123	0.069	0.071	0.075	0.081	0.088	0.095	0.103
	0.20	−0.117	−0.110	−0.091	−0.083	0.147	0.141	0.137	0.135	0.135	0.135	0.136
	0.25	−0.083	−0.079	−0.067	−0.063	0.187	0.175	0.167	0.160	0.155	0.151	0.148
	0.30	−0.064	−0.062	−0.056	−0.054	0.204	0.189	0.178	0.169	0.161	0.154	0.149
	0.40	−0.053	−0.053	−0.053	−0.053	0.207	0.190	0.177	0.165	0.156	0.148	0.141
	0.50	−0.055	−0.055	−0.056	−0.057	0.199	0.182	0.168	0.157	0.147	0.139	0.132
	0.60	−0.059	−0.059	−0.060	−0.060	0.192	0.176	0.162	0.151	0.141	0.133	0.126
	0.70	−0.061	−0.062	−0.062	−0.062	0.188	0.172	0.159	0.147	0.138	0.130	0.123
	0.80	−0.063	−0.063	−0.063	−0.063	0.186	0.170	0.157	0.146	0.137	0.129	0.122
$\dfrac{\bar\sigma_r}{\eta_1}$	0	−0.45	−0.45	−0.45	−0.495	−0.495	−0.6	−0.6	−0.6	−0.6	−0.6	0
	0.01	−0.428	−0.426	−0.426	−0.431	−0.431	−0.446	−0.453	−0.427	−0.362	−0.204	0
	0.02	−0.407	−0.403	−0.395	−0.370	−0.370	−0.347	−0.330	−0.283	−0.194	−0.059	0
	0.05	−0.344	−0.334	−0.285	−0.228	−0.227	−0.181	−0.129	−0.075	−0.027	0.000	0
	0.10	−0.248	−0.230	−0.156	−0.113	−0.113	−0.081	−0.050	−0.026	−0.010	−0.002	0
	0.15	−0.171	−0.153	−0.097	−0.076	−0.076	−0.054	−0.036	−0.023	−0.013	−0.005	0
	0.20	−0.117	−0.104	−0.073	−0.064	−0.064	−0.046	−0.033	−0.022	−0.014	−0.007	0
	0.25	−0.083	−0.076	−0.063	−0.061	−0.061	−0.045	−0.033	−0.023	−0.014	−0.007	0
	0.30	−0.064	−0.062	−0.060	−0.061	−0.061	−0.045	−0.033	−0.023	−0.015	−0.007	0
	0.40	−0.053	−0.054	−0.059	−0.062	−0.062	−0.046	−0.034	−0.024	−0.015	−0.007	0
	0.50	−0.055	−0.057	−0.061	−0.063	−0.063	−0.047	−0.034	−0.024	−0.015	−0.007	0
	0.60	−0.059	−0.060	−0.062	−0.063	−0.063	−0.047	−0.034	−0.024	−0.014	−0.007	0
	0.70	−0.061	−0.062	−0.063	−0.063	−0.063	−0.047	−0.034	−0.023	−0.014	−0.007	0
	0.80	−0.063	−0.063	−0.063	−0.063	−0.063	−0.47	−0.034	−0.023	−0.014	−0.007	0
$\dfrac{\bar\tau_{rz}}{\eta_1}$	0	0	0.077	0.184	$+\infty$	$+\infty$	0.262	0.294	0.321	0.346	0.368	0
	0.01	0	0.076	0.185	0.291	0.291	0.276	0.289	0.306	0.310	0.256	0
	0.02	0	0.076	0.187	0.264	0.264	0.278	0.273	0.267	0.234	0.126	0
	0.05	0	0.070	0.163	0.184	0.184	0.190	0.168	0.126	0.068	0.012	0
	0.10	0	0.049	0.081	0.067	0.067	0.049	0.028	0.007	−0.009	−0.013	0
	0.15	0	0.023	0.019	0.003	0.003	−0.013	−0.024	−0.030	−0.029	−0.019	0
	0.20	0	0.003	−0.014	−0.025	−0.025	−0.034	−0.039	−0.038	−0.031	−0.019	0
	0.25	0	−0.009	−0.026	−0.032	−0.032	−0.037	−0.038	−0.034	−0.027	−0.016	0
	0.30	0	−0.014	−0.028	−0.030	−0.030	−0.032	−0.031	−0.027	−0.021	−0.012	0
	0.40	0	−0.013	−0.020	−0.019	−0.019	−0.019	−0.017	−0.014	−0.010	−0.006	0
	0.50	0	−0.007	−0.010	−0.009	−0.009	−0.009	−0.007	−0.006	−0.004	−0.002	0
	0.60	0	−0.003	−0.004	−0.004	−0.004	−0.003	−0.003	−0.002	−0.001	−0.001	0
	0.70	0	−0.001	−0.001	−0.001	−0.001	−0.001	−0.001	−0.001	0.000	0.000	0
	0.80	0	0.000	0.000	0.000	0.000	0.000	0.000	0.000	0.000	0.000	0

γ = 1,5 ; μ = 0,75 ; δ = 0,5

	λ	Core					Shell					
	ρ	0	0.3	0.6	0.7	0.75	0.75	0.80	0.85	0.90	0.95	1.00
$\bar{\sigma}_z / \eta_1$	0	0	0	0	0	−0.113	0.172	0	0	0	0	0.6
	0.01	−0.015	−0.016	−0.023	−0.041	−0.140	0.104	−0.024	−0.026	−0.019	0.020	0.593
	0.02	−0.030	−0.033	−0.047	−0.084	−0.163	0.044	−0.030	−0.034	−0.004	0.119	0.577
	0.05	−0.074	−0.079	−0.114	−0.161	−0.198	−0.057	−0.023	0.027	0.125	0.297	0.534
	0.10	−0.133	−0.140	−0.167	−0.177	−0.176	−0.034	0.041	0.126	0.228	0.346	0.475
	0.15	−0.171	−0.173	−0.169	−0.154	−0.140	0.032	0.095	0.164	0.239	0.318	0.400
	0.20	−0.185	−0.181	−0.154	−0.131	−0.115	0.080	0.125	0.173	0.223	0.274	0.323
	0.25	−0.182	−0.174	−0.137	−0.114	−0.101	0.109	0.138	0.169	0.201	0.231	0.260
	0.30	−0.170	−0.159	−0.123	−0.104	−0.094	0.123	0.142	0.161	0.180	0.197	0.213
	0.40	−0.139	−0.130	−0.106	−0.096	−0.090	0.130	0.137	0.143	0.148	0.153	0.156
	0.50	−0.116	−0.111	−0.098	−0.094	−0.091	0.128	0.130	0.131	0.132	0.133	0.132
	0.60	−0.103	−0.101	−0.095	−0.094	−0.093	0.125	0.125	0.125	0.125	0.124	0.124
	0.70	−0.097	−0.096	−0.095	−0.094	−0.094	0.124	0.123	0.123	0.122	0.122	0.121
	0.80	−0.095	−0.095	−0.095	−0.095	−0.095	0.122	0.122	0.122	0.122	0.121	0.121
$\bar{\sigma}_\theta / \eta_1$	0	−0.45	−0.45	−0.45	−0.45	−0.476	−0.554	−0.6	−0.6	−0.6	−0.6	−0.4
	0.01	−0.428	−0.427	−0.424	−0.427	−0.442	−0.501	−0.516	−0.509	−0.491	−0.447	−0.299
	0.02	−0.405	−0.403	−0.398	−0.400	−0.405	−0.441	−0.439	−0.423	−0.393	−0.335	−0.233
	0.05	−0.340	−0.335	−0.320	−0.310	−0.302	−0.265	−0.244	−0.217	−0.182	−0.139	−0.093
	0.10	−0.241	−0.234	−0.207	−0.190	−0.179	−0.040	−0.025	−0.009	0.010	0.029	0.049
	0.15	−0.164	−0.156	−0.130	−0.116	−0.108	0.095	0.098	0.103	0.109	0.115	0.123
	0.20	−0.109	−0.103	−0.084	−0.076	−0.071	0.169	0.163	0.159	0.157	0.156	0.156
	0.25	−0.076	−0.071	−0.060	−0.056	−0.053	0.204	0.193	0.185	0.178	0.173	0.169
	0.30	−0.057	−0.055	−0.049	−0.047	−0.046	0.219	0.205	0.194	0.185	0.177	0.170
	0.40	−0.046	−0.045	−0.045	−0.045	−0.045	0.221	0.205	0.191	0.180	0.170	0.162
	0.50	−0.047	−0.048	−0.049	−0.049	−0.049	0.213	0.196	0.183	0.171	0.162	0.153
	0.60	−0.051	−0.051	−0.052	−0.052	−0.053	0.206	0.190	0.177	0.166	0.156	0.148
	0.70	−0.054	−0.054	−0.054	−0.054	−0.055	0.202	0.186	0.173	0.162	0.153	0.145
	0.80	−0.056	−0.056	−0.056	−0.056	−0.056	0.200	0.184	0.171	0.160	0.152	0.144
$\bar{\sigma}_r / \eta_1$	0.	−0.45	−0.45	−0.45	−0.45	−0.495	−0.495	−0.6	−0.6	−0.6	−0.6	0
	0.01	−0.428	−0.425	−0.419	−0.426	−0.419	−0.419	−0.426	−0.418	−0.359	−0.203	0
	0.02	−0.405	−0.401	−0.386	−0.377	−0.348	−0.348	−0.313	−0.273	−0.189	−0.058	0
	0.05	−0.340	−0.329	−0.281	−0.233	−0.193	−0.193	−0.139	−0.079	−0.028	−0.000	0
	0.10	−0.241	−0.224	−0.154	−0.112	−0.089	−0.089	−0.058	−0.032	−0.014	−0.004	0
	0.15	−0.164	−0.147	−0.093	−0.071	−0.061	−0.061	−0.041	−0.026	−0.015	−0.007	0
	0.20	−0.109	−0.097	−0.067	−0.058	−0.055	−0.055	−0.038	−0.026	−0.016	−0.008	0
	0.25	−0.076	−0.069	−0.056	−0.054	−0.054	−0.054	−0.039	−0.027	−0.017	−0.008	0
	0.30	−0.057	−0.055	−0.052	−0.053	−0.054	−0.054	−0.039	−0.027	−0.017	−0.008	0
	0.40	−0.046	−0.047	−0.052	−0.054	−0.055	−0.055	−0.040	−0.028	−0.017	−0.008	0
	0.50	−0.047	−0.049	−0.053	−0.055	−0.056	−0.056	−0.041	−0.028	−0.017	−0.008	0
	0.60	−0.051	−0.052	−0.055	−0.055	−0.056	−0.056	−0.040	−0.028	−0.017	−0.008	0
	0.70	−0.054	−0.054	−0.055	−0.056	−0.056	−0.056	−0.040	−0.028	−0.017	−0.008	0
	0.80	−0.056	−0.056	−0.056	−0.056	−0.056	−0.056	−0.040	−0.028	−0.017	−0.008	0
$\bar{\tau}_{rz} / \eta_1$	0	0	0.075	0.175	0.234	$+\infty$	$+\infty$	0.289	0.321	0.347	0.368	0
	0.01	0	0.075	0.175	0.241	0.309	0.309	0.299	0.309	0.313	0.257	0
	0.02	0	0.074	0.174	0.240	0.280	0.280	0.291	0.276	0.239	0.128	0
	0.05	0	0.067	0.151	0.181	0.182	0.182	0.174	0.136	0.076	0.017	0
	0.10	0	0.046	0.078	0.069	0.056	0.056	0.032	0.010	−0.007	−0.012	0
	0.15	0	0.021	0.018	0.005	−0.004	−0.004	−0.020	−0.028	−0.028	−0.020	0
	0.20	0	0.002	−0.014	−0.023	−0.027	−0.027	−0.035	−0.036	−0.031	−0.019	0
	0.25	0	−0.010	−0.026	−0.031	−0.033	−0.033	−0.035	−0.033	−0.026	−0.015	0
	0.30	0	−0.015	−0.028	−0.030	−0.030	−0.030	−0.029	−0.026	−0.020	−0.011	0
	0.40	0	−0.013	−0.020	−0.019	−0.018	−0.018	−0.017	−0.014	−0.010	−0.005	0
	0.50	0	−0.008	−0.010	−0.010	−0.009	−0.009	−0.008	−0.006	−0.004	−0.002	0
	0.60	0	−0.004	−0.005	−0.004	−0.004	−0.004	−0.003	−0.002	−0.001	−0.001	0
	0.70	0	−0.001	−0.002	−0.001	−0.001	−0.001	−0.001	−0.001	0.000	0.000	0
	0.80	0	0.000	0.000	0.000	0.000	0.000	0.000	0.000	0.000	0.000	0

	λ / p	Core						Shell				
		0	0.3	0.6	0.7	0.75	0.80	0.80	0.85	0.90	0.95	1.00
$\dfrac{\bar\sigma_z}{\eta_1}$	0	0	0	0	0	0	−0.113	0.172	0	0	0	0.6
	0.01	−0.015	−0.016	−0.022	−0.029	−0.042	−0.143	0.095	−0.024	−0.019	0.021	0.593
	0.02	−0.030	−0.032	−0.043	−0.059	−0.086	−0.166	0.031	−0.022	−0.002	0.122	0.579
	0.05	−0.072	−0.077	−0.101	−0.129	−0.155	−0.185	−0.040	0.031	0.135	0.310	0.553
	0.10	−0.130	−0.135	−0.151	−0.154	−0.150	−0.140	0.033	0.128	0.239	0.365	0.504
	0.15	−0.165	−0.165	−0.154	−0.136	−0.122	−0.102	0.106	0.177	0.256	0.339	0.424
	0.20	−0.177	−0.171	−0.140	−0.116	−0.100	−0.081	0.147	0.194	0.245	0.296	0.346
	0.25	−0.172	−0.163	−0.124	−0.101	−0.087	−0.073	0.164	0.194	0.225	0.256	0.284
	0.30	−0.159	−0.148	−0.111	−0.092	−0.081	−0.070	0.168	0.187	0.205	0.223	0.238
	0.40	−0.127	−0.118	−0.094	−0.084	−0.078	−0.073	0.163	0.169	0.175	0.179	0.183
	0.50	−0.103	−0.099	−0.086	−0.081	−0.079	−0.077	0.155	0.157	0.158	0.158	0.158
	0.60	−0.091	−0.088	−0.083	−0.081	−0.080	−0.080	0.150	0.150	0.150	0.149	0.149
	0.70	−0.085	−0.084	−0.082	−0.082	−0.081	−0.081	0.148	0.147	0.147	0.146	0.146
	0.80	−0.082	−0.082	−0.082	−0.082	−0.082	−0.082	0.146	0.146	0.146	0.146	0.146
$\dfrac{\bar\sigma_\theta}{\eta_1}$	0	−0.45	−0.45	−0.45	−0.45	−0.45	−0.476	−0.554	−0.6	−0.6	−0.6	−0.4
	0.01	−0.427	−0.426	−0.422	−0.421	−0.422	−0.437	−0.493	−0.504	−0.488	−0.445	−0.297
	0.02	−0.403	−0.401	−0.394	−0.391	−0.392	−0.395	−0.426	−0.416	−0.387	−0.330	−0.229
	0.05	−0.336	−0.331	−0.313	−0.301	−0.293	−0.282	−0.234	−0.205	−0.170	−0.127	−0.080
	0.10	−0.235	−0.227	−0.198	−0.181	−0.171	−0.159	−0.005	0.012	0.031	0.050	0.071
	0.15	−0.156	−0.148	−0.121	−0.108	−0.100	−0.092	0.125	0.128	0.133	0.139	0.146
	0.20	−0.101	−0.094	−0.076	−0.067	−0.063	−0.058	0.192	0.187	0.183	0.181	0.180
	0.25	−0.067	−0.063	−0.052	−0.047	−0.045	−0.043	0.224	0.213	0.205	0.198	0.192
	0.30	−0.048	−0.046	−0.040	−0.038	−0.037	−0.036	0.236	0.223	0.212	0.202	0.194
	0.40	−0.037	−0.037	−0.036	−0.036	−0.036	−0.037	0.236	0.221	0.208	0.197	0.187
	0.50	−0.038	−0.039	−0.040	−0.040	−0.040	−0.041	0.228	0.212	0.199	0.188	0.178
	0.60	−0.042	−0.042	−0.043	−0.044	−0.044	−0.044	0.221	0.206	0.193	0.182	0.173
	0.70	−0.045	−0.045	−0.045	−0.046	−0.046	−0.046	0.218	0.203	0.190	0.179	0.170
	0.80	−0.047	−0.047	−0.047	−0.047	−0.047	−0.047	0.215	0.200	0.188	0.177	0.168
$\dfrac{\bar\sigma_r}{\eta_1}$	0	−0.45	−0.45	−0.45	−0.45	−0.45	−0.495	−0.495	−0.6	−0.6	−0.6	0
	0.01	−0.427	−0.424	−0.414	−0.411	−0.414	−0.401	−0.401	−0.394	−0.352	−0.201	0
	0.02	−0.403	−0.399	−0.379	−0.367	−0.355	−0.317	−0.317	−0.260	−0.182	−0.056	0
	0.05	−0.336	−0.324	−0.275	−0.233	−0.198	−0.152	−0.152	−0.091	−0.034	−0.002	0
	0.10	−0.235	−0.217	−0.149	−0.109	−0.087	−0.065	−0.065	−0.038	−0.018	−0.006	0
	0.15	−0.156	−0.138	−0.087	−0.065	−0.056	−0.047	−0.047	−0.030	−0.017	−0.008	0
	0.20	−0.101	−0.089	−0.060	−0.051	−0.048	−0.045	−0.045	−0.030	−0.018	−0.009	0
	0.25	−0.067	−0.061	−0.049	−0.046	−0.046	−0.046	−0.046	−0.031	−0.020	−0.010	0
	0.30	−0.048	−0.046	−0.044	−0.045	−0.046	−0.047	−0.047	−0.032	−0.020	−0.010	0
	0.40	−0.037	−0.038	−0.043	−0.045	−0.046	−0.047	−0.047	−0.033	−0.020	−0.010	0
	0.50	−0.038	−0.040	−0.045	−0.046	−0.047	−0.047	−0.047	−0.033	−0.020	−0.009	0
	0.60	−0.042	−0.043	−0.046	−0.047	−0.047	−0.047	−0.047	−0.033	−0.020	−0.009	0
	0.70	−0.045	−0.045	−0.047	−0.047	−0.047	−0.047	−0.047	−0.032	−0.020	−0.009	0
	0.80	−0.047	−0.047	−0.047	−0.047	−0.047	−0.047	−0.047	−0.032	−0.020	−0.009	0
$\dfrac{\bar\tau_{rz}}{\eta_1}$	0	0	0.074	0.169	0.217	0.252	$+\infty$	$+\infty$	0.318	0.349	0.370	0
	0.01	0	0.074	0.168	0.217	0.258	0.330	0.330	0.320	0.317	0.260	0
	0.02	0	0.073	0.165	0.214	0.254	0.291	0.291	0.295	0.249	0.133	0
	0.05	0	0.065	0.141	0.170	0.178	0.171	0.171	0.143	0.086	0.023	0
	0.10	0	0.044	0.073	0.067	0.056	0.041	0.041	0.013	−0.006	−0.012	0
	0.15	0	0.019	0.016	0.004	−0.003	−0.012	−0.012	−0.025	−0.028	−0.020	0
	0.20	0	0.000	−0.015	−0.023	−0.027	−0.030	−0.030	−0.033	−0.029	−0.019	0
	0.25	0	−0.011	−0.027	−0.031	−0.032	−0.032	−0.032	−0.031	−0.025	−0.015	0
	0.30	0	−0.015	−0.028	−0.030	−0.029	−0.028	−0.028	−0.025	−0.020	−0.011	0
	0.40	0	−0.013	−0.020	−0.019	−0.018	−0.017	−0.017	−0.014	−0.010	−0.006	0
	0.50	0	−0.008	−0.011	−0.010	−0.009	−0.008	−0.008	−0.007	−0.005	−0.002	0
	0.60	0	−0.004	−0.005	−0.004	−0.004	−0.004	−0.004	−0.003	−0.002	−0.001	0
	0.70	0	−0.001	−0.002	−0.002	−0.001	−0.001	−0.001	−0.001	−0.001	0.000	0
	0.80	0	0.000	0.000	0.000	0.000	0.000	0.000	0.000	0.000	0.000	0

$\gamma = 1,5 ; \mu = 0,85 ; \delta = 0,5$

	λ	Core							Shell			
	p	0	0.3	0.6	0.7	0.75	0.80	0.85	0.85	0.90	0.95	1.00
$\dfrac{\bar{\sigma}_z}{\eta_1}$	0	0	0	0	0	0	0	-0.113	0.172	0	0	0.6
	0.01	-0.015	-0.016	-0.020	-0.025	-0.030	-0.043	-0.146	0.082	-0.017	0.023	0.594
	0.02	-0.029	-0.031	-0.040	-0.049	-0.059	-0.087	-0.167	0.019	0.009	0.129	0.585
	0.05	-0.071	-0.075	-0.092	-0.107	-0.121	-0.139	-0.156	0.010	0.141	0.329	0.587
	0.10	-0.127	-0.130	-0.135	-0.132	-0.125	-0.111	-0.087	0.134	0.253	0.390	0.539
	0.15	-0.159	-0.157	-0.137	-0.116	-0.100	-0.079	-0.052	0.203	0.282	0.367	0.454
	0.20	-0.168	-0.161	-0.124	-0.098	-0.081	-0.062	-0.040	0.227	0.276	0.328	0.378
	0.25	-0.162	-0.150	-0.108	-0.084	-0.071	-0.056	-0.040	0.228	0.258	0.289	0.317
	0.30	-0.147	-0.135	-0.095	-0.076	-0.066	-0.055	-0.044	0.220	0.239	0.256	0.272
	0.40	-0.113	-0.103	-0.079	-0.068	-0.063	-0.058	-0.053	0.201	0.207	0.211	0.215
	0.50	-0.089	-0.084	-0.071	-0.066	-0.064	-0.062	-0.060	0.187	0.188	0.189	0.189
	0.60	-0.076	-0.073	-0.068	-0.066	-0.065	-0.065	-0.064	0.179	0.179	0.179	0.178
	0.70	-0.070	-0.069	-0.067	-0.066	-0.066	-0.066	-0.066	0.176	0.175	0.175	0.174
	0.80	-0.067	-0.067	-0.067	-0.067	-0.067	-0.067	-0.067	0.174	0.174	0.174	0.174
$\dfrac{\bar{\sigma}_\theta}{\eta_1}$	0	-0.45	-0.45	-0.45	-0.45	-0.45	-0.45	-0.476	-0.554	-0.6	-0.6	-0.4
	0.01	-0.426	-0.425	-0.420	-0.418	-0.416	-0.417	-0.428	-0.481	-0.482	-0.441	-0.294
	0.02	-0.402	-0.399	-0.390	-0.385	-0.382	-0.380	-0.379	-0.404	-0.378	-0.323	-0.223
	0.05	-0.330	-0.326	-0.305	-0.292	-0.284	-0.272	-0.257	-0.193	-0.155	-0.110	-0.061
	0.10	-0.227	-0.218	-0.189	-0.172	-0.161	-0.149	-0.135	0.038	0.056	0.076	0.097
	0.15	-0.146	-0.138	-0.111	-0.098	-0.090	-0.082	-0.074	0.159	0.162	0.167	0.173
	0.20	-0.090	-0.084	-0.065	-0.057	-0.053	-0.048	-0.044	0.220	0.214	0.210	0.207
	0.25	-0.056	-0.052	-0.041	-0.037	-0.035	-0.032	-0.030	0.247	0.236	0.228	0.221
	0.30	-0.038	-0.036	-0.030	-0.028	-0.027	-0.026	-0.025	0.257	0.243	0.232	0.223
	0.40	-0.026	-0.026	-0.026	-0.026	-0.026	-0.026	-0.027	0.255	0.239	0.226	0.215
	0.50	-0.028	-0.029	-0.030	-0.030	-0.030	-0.031	-0.031	0.246	0.231	0.218	0.206
	0.60	-0.032	-0.032	-0.033	-0.034	-0.034	-0.034	-0.034	0.239	0.224	0.211	0.200
	0.70	-0.035	-0.035	-0.036	-0.036	-0.036	-0.036	-0.036	0.235	0.221	0.208	0.197
	0.80	-0.037	-0.037	-0.037	-0.037	-0.037	-0.037	-0.037	0.233	0.218	0.206	0.195
$\dfrac{\bar{\sigma}_r}{\eta_1}$	0	-0.45	-0.45	-0.45	-0.45	-0.45	-0.45	-0.495	-0.495	-0.6	-0.6	0
	0.01	-0.426	-0.423	-0.411	-0.403	-0.399	-0.397	-0.373	-0.373	-0.331	-0.197	0
	0.02	-0.402	-0.396	-0.373	-0.356	-0.344	-0.324	-0.270	-0.270	-0.175	-0.053	0
	0.05	-0.330	-0.319	-0.267	-0.227	-0.197	-0.156	-0.104	-0.104	-0.045	-0.007	0
	0.10	-0.227	-0.208	-0.141	-0.103	-0.082	-0.061	-0.042	-0.042	-0.020	-0.007	0
	0.15	-0.146	-0.129	-0.079	-0.058	-0.049	-0.041	-0.035	-0.035	-0.020	-0.009	0
	0.20	-0.090	-0.079	-0.051	-0.043	-0.040	-0.037	-0.036	-0.036	-0.022	-0.010	0
	0.25	-0.056	-0.050	-0.039	-0.037	-0.037	-0.037	-0.038	-0.038	-0.023	-0.011	0
	0.30	-0.038	-0.036	-0.035	-0.036	-0.036	-0.037	-0.038	-0.038	-0.024	-0.011	0
	0.40	-0.026	-0.028	-0.033	-0.035	-0.036	-0.038	-0.038	-0.038	-0.024	-0.011	0
	0.50	-0.028	-0.030	-0.034	-0.036	-0.037	-0.037	-0.038	-0.038	-0.023	-0.011	0
	0.60	-0.032	-0.033	-0.035	-0.037	-0.037	-0.037	-0.038	-0.038	-0.023	-0.011	0
	0.70	-0.035	-0.035	-0.036	-0.037	-0.037	-0.037	-0.038	-0.038	-0.023	-0.011	0
	0.80	-0.037	-0.037	-0.037	-0.037	-0.037	-0.037	-0.037	-0.037	-0.023	-0.011	0
$\dfrac{\bar{\tau}_{rz}}{\eta_1}$	0	0	0.073	0.163	0.205	0.233	0.271	$+\infty$	$+\infty$	0.349	0.372	0
	0.01	0	0.072	0.162	0.204	0.232	0.274	0.346	0.346	0.331	0.265	0
	0.02	0	0.071	0.158	0.198	0.225	0.262	0.295	0.295	0.270	0.144	0
	0.05	0	0.063	0.132	0.156	0.164	0.165	0.146	0.146	0.093	0.030	0
	0.10	0	0.041	0.066	0.061	0.052	0.040	0.023	0.023	-0.003	-0.013	0
	0.15	0	0.016	0.012	0.001	-0.005	-0.012	-0.019	-0.019	-0.025	-0.019	0
	0.20	0	-0.002	-0.017	-0.024	-0.027	-0.030	-0.030	-0.030	-0.028	-0.018	0
	0.25	0	-0.012	-0.028	-0.031	-0.032	-0.032	-0.030	-0.030	-0.024	-0.015	0
	0.30	0	-0.016	-0.029	-0.030	-0.029	-0.028	-0.026	-0.026	-0.020	-0.011	0
	0.40	0	-0.013	-0.020	-0.019	-0.018	-0.017	-0.015	-0.015	-0.011	-0.006	0
	0.50	0	-0.008	-0.011	-0.010	-0.009	-0.008	-0.007	-0.007	-0.005	-0.003	0
	0.60	0	-0.004	-0.005	-0.004	-0.004	-0.004	-0.003	-0.003	-0.002	-0.001	0
	0.70	0	-0.002	-0.002	-0.002	-0.001	-0.001	-0.001	-0.001	-0.001	0.000	0
	0.80	0	0.000	0.000	0.000	0.000	0.000	0.000	0.000	0.000	0.000	0

	λ \ p	Core								Shell		
		0	**0.3**	**0.6**	**0.7**	**0.75**	**0.80**	**0.85**	**0.90**	**0.90**	**0.95**	**1.00**
$\dfrac{\bar{\sigma}^z}{\eta_1}$	0	0	0	0	0	0	0	0	−0.113	0.172	0	0.6
	0.01	−0.015	−0.015	−0.019	−0.022	−0.025	−0.030	−0.044	−0.151	0.064	0.027	0.598
	0.02	−0.029	−0.031	−0.037	−0.043	−0.048	−0.058	−0.084	−0.158	0.023	0.147	0.607
	0.05	−0.070	−0.073	−0.084	−0.091	−0.096	−0.101	−0.103	−0.089	0.132	0.354	0.644
	0.10	−0.123	−0.125	−0.121	−0.109	−0.098	−0.079	−0.051	−0.009	0.286	0.429	0.585
	0.15	−0.152	−0.148	−0.120	−0.094	−0.075	−0.051	−0.022	0.012	0.329	0.413	0.500
	0.20	−0.158	−0.149	−0.104	−0.076	−0.058	−0.037	−0.015	0.010	0.325	0.375	0.426
	0.25	−0.149	−0.136	−0.089	−0.064	−0.049	−0.034	−0.017	−0.001	0.304	0.335	0.364
	0.30	−0.132	−0.119	−0.077	−0.056	−0.045	−0.034	−0.023	−0.012	0.281	0.300	0.316
	0.40	−0.096	−0.087	−0.060	−0.050	−0.044	−0.039	−0.034	−0.030	0.246	0.251	0.254
	0.50	−0.071	−0.066	−0.053	−0.048	−0.046	−0.044	−0.042	−0.041	0.224	0.225	0.225
	0.60	−0.058	−0.055	−0.050	−0.048	−0.047	−0.047	−0.046	−0.046	0.214	0.213	0.212
	0.70	−0.052	−0.051	−0.049	−0.048	−0.048	−0.048	−0.048	−0.048	0.210	0.209	0.208
	0.80	−0.049	−0.049	−0.049	−0.049	−0.049	−0.049	−0.049	−0.049	0.208	0.208	0.208
$\dfrac{\bar{\sigma}^\theta}{\eta_1}$	0	−0.45	−0.45	−0.45	−0.45	−0.45	−0.45	−0.45	−0.476	−0.554	−0.6	−0.4
	0.01	−0.425	−0.423	−0.418	−0.415	−0.412	−0.410	−0.408	−0.413	−0.463	−0.434	−0.289
	0.02	−0.399	−0.397	−0.387	−0.380	−0.375	−0.370	−0.363	−0.354	−0.369	−0.312	−0.211
	0.05	−0.326	−0.320	−0.297	−0.282	−0.273	−0.260	−0.244	−0.223	−0.135	−0.086	−0.033
	0.10	−0.217	−0.209	−0.177	−0.159	−0.148	−0.136	−0.123	−0.108	0.089	0.108	0.128
	0.15	−0.133	−0.125	−0.098	−0.085	−0.077	−0.069	−0.061	−0.053	0.199	0.202	0.205
	0.20	−0.077	−0.071	−0.053	−0.044	−0.040	−0.036	−0.031	−0.027	0.252	0.246	0.241
	0.25	−0.043	−0.039	−0.028	−0.024	−0.022	−0.020	−0.018	−0.016	0.274	0.263	0.254
	0.30	−0.025	−0.023	−0.018	−0.016	−0.015	−0.014	−0.013	−0.012	0.281	0.268	0.257
	0.40	−0.014	−0.014	−0.014	−0.014	−0.014	−0.015	−0.015	−0.015	0.276	0.261	0.248
	0.50	−0.017	−0.017	−0.018	−0.019	−0.019	−0.019	−0.019	−0.020	0.266	0.251	0.238
	0.60	−0.021	−0.021	−0.022	−0.022	−0.023	−0.023	−0.023	−0.023	0.259	0.244	0.231
	0.70	−0.024	−0.024	−0.024	−0.025	−0.025	−0.025	−0.025	−0.025	0.255	0.240	0.228
	0.80	−0.026	−0.026	−0.026	−0.026	−0.026	−0.026	−0.026	−0.026	0.252	0.238	0.227
$\dfrac{\bar{\sigma}^r}{\eta_1}$	0	−0.45	−0.45	−0.45	−0.45	−0.45	−0.45	−0.45	−0.495	−0.495	−0.6	0
	0.01	−0.425	−0.422	−0.408	−0.398	−0.391	−0.381	−0.368	−0.319	−0.319	−0.184	0
	0.02	−0.399	−0.393	−0.368	−0.348	−0.333	−0.312	−0.276	−0.194	−0.194	−0.054	0
	0.05	−0.326	−0.312	−0.257	−0.218	−0.189	−0.152	−0.106	−0.054	−0.054	−0.012	0
	0.10	−0.217	−0.198	−0.131	−0.094	−0.075	−0.055	−0.037	−0.024	−0.024	−0.008	0
	0.15	−0.133	−0.116	−0.068	−0.049	−0.040	−0.033	−0.028	−0.025	−0.025	−0.011	0
	0.20	−0.077	−0.066	−0.040	−0.033	−0.030	−0.028	−0.027	−0.027	−0.027	−0.012	0
	0.25	−0.043	−0.038	−0.028	−0.027	−0.026	−0.026	−0.027	−0.027	−0.027	−0.013	0
	0.30	−0.025	−0.023	−0.023	−0.024	−0.025	−0.026	−0.027	−0.028	−0.028	−0.013	0
	0.40	−0.014	−0.016	−0.022	−0.024	−0.025	−0.026	−0.027	−0.028	−0.028	−0,013	0
	0.50	−0.017	−0.019	−0.023	−0.025	−0.025	−0.026	−0.026	−0.027	−0.027	−0.013	0
	0.60	−0.021	−0.022	−0.025	−0.026	−0.026	−0.026	−0.026	−0.027	−0.027	−0.012	0
	0.70	−0.024	−0.024	−0.026	−0.026	−0.026	−0.026	−0.026	−0.027	−0.027	−0.012	0
	0.80	−0.026	−0.026	−0.026	−0.026	−0.026	−0.026	−0.026	−0.026	−0.026	−0.012	0
$\dfrac{\bar{\tau}^{rz}}{\eta_1}$	0	0	0.071	0.157	0.196	0.219	0.248	0.287	$+\infty$	$+\infty$	0.377	0
	0.01	0	0.071	0.156	0.194	0.216	0.244	0.284	0.354	0.354	0.283	0
	0.02	0	0.070	0.152	0.186	0.207	0.230	0.260	0.276	0.276	0.166	0
	0.05	0	0.061	0.124	0.142	0.148	0.147	0.136	0.103	0.103	0.032	0
	0.10	0	0.038	0.058	0.051	0.044	0.032	0.018	0.003	0.003	−0.012	0
	0.15	0	0.013	0.006	−0.004	−0.010	−0.016	−0.021	−0.023	−0.023	−0.018	0
	0.20	0	−0.005	−0.021	−0.027	−0.030	−0.031	−0.031	−0.028	−0.028	−0.018	0
	0.25	0	−0.015	−0.030	−0.033	−0.033	−0.032	−0.030	−0.026	−0.026	−0.015	0
	0.30	0	−0.018	−0.030	−0.031	−0.030	−0.028	−0.026	−0.021	−0.021	−0.012	0
	0.40	0	−0.014	−0.021	−0.020	−0.019	−0.017	−0.015	−0.012	−0.012	−0.006	0
	0.50	0	−0.008	−0.011	−0.010	−0.009	−0.008	−0.007	−0.006	−0.006	−0.003	0
	0.60	0	−0.004	−0.007	−0.005	−0.004	−0.004	−0.003	−0.003	−0.003	−0.001	0
	0.70	0	−0.002	−0.003	−0.002	−0.001	−0.001	−0.001	0.000	0.000	0.000	0
	0.80	0	0.000	0.000	0.000	0.000	0.000	0.000	0.000	0.000	0.000	0

$\gamma = 1,5 \; ; \; \mu = 0,95 \; ; \; \delta = 0,5$

	λ	Core									Shell	
	p	0	0.3	0.6	0.7	**0.75**	0.80	0.85	**0.90**	0.95	0.95	1.00
$\dfrac{\bar{\sigma}_z}{\eta_1}$	0	0	0	0	0	0	0	0	0	−0.113	**0.172**	0.6
	0.01	−0.014	−0.015	−0.018	−0.020	−0.021	−0.024	−0.028	−0.041	−0.139	0.062	0.628
	0.02	−0.029	−0.030	−0.035	−0.038	−0.041	−0.044	−0.050	−0.063	−0.084	0.149	0.694
	0.05	−0.069	−0.071	−0.077	−0.078	−0.075	−0.069	−0.055	−0.021	0.053	0.409	0.735
	0.10	−0.120	−0.120	−0.106	−0.085	−0.068	−0.042	−0.005	0.044	0.106	0.514	0.672
	0.15	−0.146	−0.139	−0.099	−0.066	−0.044	−0.016	0.017	0.055	0.096	0.496	0.585
	0.20	−0.147	−0.135	−0.082	−0.049	−0.028	−0.005	0.019	0.046	0.073	0.448	0.501
	0.25	−0.135	−0.119	−0.066	−0.037	−0.021	−0.004	0.013	0.031	0.048	0.399	0.429
	0.30	−0.114	−0.099	−0.053	−0.031	−0.019	−0.007	0.005	0.016	0.027	0.356	0.373
	0.40	−0.075	−0.065	−0.037	−0.027	−0.021	−0.015	−0.011	−0.006	−0.003	0.297	0.301
	0.50	−0.048	−0.043	−0.030	−0.025	−0.023	−0.021	−0.020	−0.019	−0.018	0.267	0.281
	0.60	−0.035	−0.033	−0.028	−0.026	−0.025	−0.025	−0.024	−0.024	−0.024	0.254	0.259
	0.70	−0.029	−0.028	−0.027	−0.026	−0.026	−0.026	−0.026	−0.026	−0.027	0.250	0.251
	0.80	−0.027	−0.027	−0.027	−0.027	−0.027	−0.027	−0.027	−0.027	−0.027	0.249	0.249
$\dfrac{\bar{\sigma}_\theta}{\eta_1}$	0	−0.45	−0.45	−0.45	−0.45	−0.45	−0.45	−0.45	−0.45	−0.476	−0.554	−0.4
	0.01	−0.423	−0.422	−0.415	−0.412	−0.409	−0.404	−0.400	−0.391	−0.381	−0.422	−0.276
	0.02	−0.396	−0.394	−0.383	−0.374	−0.369	−0.361	−0.351	−0.335	−0.307	−0.298	−0.182
	0.05	−0.319	−0.312	−0.287	−0.271	−0.260	−0.246	−0.228	−0.205	−0.176	−0.050	0.008
	0.10	−0.205	−0.195	−0.162	−0.143	−0.131	−0.119	−0.105	−0.090	−0.074	0.154	0.173
	0.15	−0.119	−0.110	−0.081	−0.067	−0.060	−0.052	−0.043	−0.035	−0.027	0.249	0.251
	0.20	−0.060	−0.054	−0.035	−0.027	−0.023	−0.018	−0.014	−0.010	−0.006	0.292	0.285
	0.25	−0.026	−0.022	−0.012	−0.008	−0.005	−0.003	−0.001	0.001	0.002	0.308	0.297
	0.30	−0.008	−0.007	−0.002	0.000	0.001	0.002	0.002	0.003	0.003	0.310	0.297
	0.40	0.000	0.000	0.000	0.000	0.000	−0.001	−0.001	−0.001	−0.002	0.300	0.285
	0.50	−0.003	−0.004	−0.005	−0.006	−0.006	−0.006	−0.007	−0.007	−0.007	0.288	0.274
	0.60	−0.008	−0.009	−0.010	−0.010	−0.010	−0.010	−0.011	−0.010	−0.011	0.281	0.266
	0.70	−0.011	−0.012	−0.012	−0.012	−0.012	−0.013	−0.013	−0.013	−0.013	0.277	0.263
	0.80	−0.014	−0.014	−0.014	−0.014	−0.014	−0.014	−0.014	−0.014	−0.014	0.275	0.261
$\dfrac{\bar{\sigma}_r}{\eta_1}$	0	−0.45	−0.45	−0.45	−0.45	−0.45	−0.45	−0.45	−0.45	−0.495	−0.495	0
	0.01	−0.423	−0.420	−0.405	−0.394	−0.384	−0.371	−0.351	−0.315	−0.196	−0.196	0
	0.02	−0.396	−0.390	−0.362	−0.339	−0.322	−0.298	−0.261	−0.198	−0.078	−0.078	0
	0.05	−0.319	−0.305	−0.245	−0.203	−0.175	−0.139	−0.097	−0.051	−0.014	−0.014	0
	0.10	−0.205	−0.185	−0.116	−0.080	−0.062	−0.044	−0.028	−0.017	−0.011	−0.011	0
	0.15	−0.119	−0.101	−0.053	−0.035	−0.028	−0.021	−0.017	−0.014	−0.013	−0.013	0
	0.20	−0.060	−0.050	−0.025	−0.019	−0.017	−0.015	−0.015	−0.015	−0.015	−0.015	0
	0.25	−0.026	−0.022	−0.014	−0.013	−0.013	−0.013	−0.014	−0.015	−0.015	−0.015	0
	0.30	−0.008	−0.008	−0.009	−0.011	−0.012	−0.013	−0.014	−0.015	−0.015	−0.015	0
	0.40	0.000	−0.002	−0.008	−0.011	−0.012	−0.013	−0.014	−0.015	−0.015	−0.015	0
	0.50	−0.003	−0.005	−0.010	−0.012	−0.013	−0.013	−0.014	−0.014	−0.015	−0.015	0
	0.60	−0.008	−0.009	−0.012	−0.013	−0.013	−0.014	−0.014	−0.014	−0.014	−0.014	0
	0.70	−0.011	−0.012	−0.013	−0.014	−0.014	−0.014	−0.014	−0.014	−0.014	−0.014	0
	0.80	−0.014	−0.014	−0.014	−0.014	−0.014	−0.014	−0.014	−0.014	−0.014	−0.014	0
$\dfrac{\bar{\tau}_{rz}}{\eta_1}$	0	0	0.070	0.153	0.188	0.207	0.230	0.257	0.297	$+\infty$	$+\infty$	0
	0.01	0	0.070	0.152	0.184	0.203	0.224	0.248	0.281	0.314	0.314	0
	0.02	0	0.069	0.146	0.176	0.192	0.207	0.222	0.231	0.187	0.187	0
	0.05	0	0.060	0.116	0.128	0.130	0.126	0.112	0.083	0.041	0.041	0
	0.10	0	0.035	0.048	0.039	0.030	0.019	0.007	−0.004	−0.010	−0.010	0
	0.15	0	0.009	−0.002	−0.013	−0.018	−0.023	−0.026	−0.026	−0.021	−0.021	0
	0.20	0	−0.009	−0.027	−0.033	−0.035	−0.035	−0.033	−0.029	−0.021	−0.021	0
	0.25	0	−0.018	−0.034	−0.037	−0.036	−0.034	−0.032	−0.025	−0.018	−0.018	0
	0.30	0	−0.020	−0.033	−0.031	−0.031	−0.029	−0.031	−0.020	−0.014	−0.014	0
	0.40	0	−0.015	−0.021	−0.020	−0.019	−0.017	−0.019	−0.011	−0.007	−0.007	0
	0.50	0	−0.008	−0.011	−0.010	−0.009	−0.008	−0.010	−0.005	−0.003	−0.003	0
	0.60	0	−0.004	−0.005	−0.004	−0.004	−0.003	−0.004	−0.002	−0.001	−0.001	0
	0.70	0	−0.001	−0.002	−0.001	−0.001	−0.001	−0.001	−0.001	0.000	0.000	0
	0.80	0	0.000	0.000	0.000	0.000	0.000	0.000	0.000	0.000	0.000	0

	λ	Core				Shell						
	ρ	0	0.8	0.6	0.7	0.7	0.75	0.80	0.85	0.90	0,95	1.00
$\bar{\sigma}_z/\eta_1$	0	0	0	0	−0.228	−0.261	0	0	0	0	0	0.6
	0.01	−0.022	−0.024	−0.047	−0.255	0.193	0.006	−0.011	−0.012	−0.007	0.032	0.608
	0.02	−0.043	−0.048	−0.095	−0.280	0.134	0.015	−0.014	−0.011	0.018	0.142	0.606
	0.05	−0.106	−0.117	−0.210	−0.323	0.015	0.014	0.026	0.072	0.171	0.347	0.592
	0.10	−0.195	−0.212	−0.288	−0.321	−0.005	0.053	0.119	0.200	0.302	0.424	0.561
	0.15	−0.256	−0.269	−0.294	−0.288	0.052	0.111	0.176	0.248	0.326	0.412	0.501
	0.20	−0.289	−0.291	−0.278	−0.258	0.110	0.155	0.205	0.259	0.314	0.372	0.429
	0.25	−0.296	−0.291	−0.260	−0.237	0.151	0.184	0.218	0.255	0.292	0.328	0.363
	0.30	−0.288	−0.280	−0.244	−0.224	0.177	0.199	0.222	0.245	0.268	0.290	0.310
	0.40	−0.259	−0.250	−0.224	−0.212	0.202	0.209	0.217	0.226	0.233	0.240	0.244
	0.50	−0.235	−0.229	−0.215	−0.210	0.206	0.209	0.211	0.213	0.214	0.215	0.216
	0.60	−0.221	−0.218	−0.212	−0.210	0.206	0.206	0.206	0.206	0.206	0.206	0.205
	0.70	−0.214	−0.213	−0.211	−0.211	0.204	0.204	0.204	0.204	0.203	0.203	0.202
	0.80	−0.211	−0.211	−0.211	−0.211	0.203	0.203	0.203	0.203	0.203	0.203	0.203
$\bar{\sigma}_\theta/\eta_1$	0	−0.6	−0.6	−0.6	−0.635	−0.554	−0.6	−0.6	−0.6	−0.6	−0.6	−0.4
	0.01	−0.575	−0.574	−0.574	−0.597	−0.491	−0.513	−0.511	−0.502	−0.484	−0.440	−0.291
	0.02	−0.550	−0.548	−0.546	−0.558	−0.423	−0.431	−0.425	−0.409	−0.378	−0.320	−0.217
	0.05	−0.475	−0.471	−0.456	−0.444	−0.226	−0.218	−0.203	−0.180	−0.147	−0.105	−0.059
	0.10	−0.362	−0.353	−0.321	−0.300	0.038	0.040	0.047	0.058	0.073	0.090	0.108
	0.15	−0.269	−0.260	−0.228	−0.211	0.207	0.197	0.192	0.191	0.192	0.195	0.200
	0.20	−0.203	−0.195	−0.171	−0.160	0.306	0.286	0.272	0.262	0.254	0.249	0.246
	0.25	−0.160	−0.155	−0.140	−0.133	0.358	0.333	0.312	0.297	0.284	0.273	0.265
	0.30	−0.135	−0.132	−0.124	−0.121	0.383	0.354	0.330	0.311	0.295	0.281	0.270
	0.40	−0.118	−0.117	−0.116	−0.116	0.393	0.360	0.334	0.312	0.293	0.278	0.264
	0.50	−0.118	−0.118	−0.119	−0.120	0.386	0.353	0.326	0.304	0.285	0.269	0.255
	0.60	−0.122	−0.122	−0.123	−0.123	0.379	0.346	0.319	0.297	0.278	0.263	0.249
	0.70	−0.125	−0.125	−0.125	−0.125	0.374	0.342	0.315	0.293	0.275	0.259	0.246
	0.80	−0.127	−0.127	−0.127	−0.127	0.371	0.339	0.313	0.291	0.273	0.257	0.244
$\bar{\sigma}_r/\eta_1$	0	−0.6	−0.6	−0.6	−0.584	−0.584	−0.6	−0.6	−0.6	−0.6	−0.6	0
	0.01	−0.575	−0.572	−0.566	−0.516	−0.516	−0.474	−0.466	−0.433	−0.365	−0.205	0
	0.02	−0.550	−0.545	−0.526	−0.459	−0.459	−0.391	−0.353	−0.295	−0.199	−0.061	0
	0.05	−0.475	−0.463	−0.393	−0.311	−0.311	−0.238	−0.165	−0.096	−0.037	−0.004	0
	0.10	−0.362	−0.339	−0.243	−0.188	−0.188	−0.135	−0.088	−0.050	−0.023	−0.008	0
	0.15	−0.269	−0.246	−0.173	−0.144	−0.144	−0.105	−0.072	−0.046	−0.026	−0.011	0
	0.20	−0.203	−0.186	−0.143	−0.130	−0.130	−0.095	−0.068	−0.045	−0.028	−0.013	0
	0.25	−0.160	−0.150	−0.130	−0.126	−0.126	−0.093	−0.067	−0.046	−0.028	−0.014	0
	0.30	−0.135	−0.131	−0.124	−0.124	−0.124	−0.093	−0.067	−0.046	−0.029	−0.014	0
	0.40	−0.118	−0.119	−0.122	−0.125	−0.125	−0.094	−0.068	−0.047	−0.029	−0.014	0
	0.50	−0.118	−0.120	−0.124	−0.126	−0.126	−0.095	−0.069	−0.047	−0.029	−0.014	0
	0.60	−0.122	−0.123	−0.126	−0.127	−0.127	−0.095	−0.069	−0.047	−0.029	−0.013	0
	0.70	−0.125	−0.125	−0.126	−0.127	−0.127	−0.095	−0.069	−0.047	−0.029	−0.013	0
	0.80	−0.127	−0.127	−0.127	−0.127	−0.127	−0.095	−0.069	−0.047	−0.029	−0.013	0
$\bar{\tau}_{rz}/\eta_1$	0	0	0.109	0.282	+∞	+∞	0.399	0.388	0.388	0.390	0.390	0
	0.01	0	0.109	0.283	0.458	0.458	0.404	0.381	0.371	0.353	0.278	0
	0.02	0	0.108	0.281	0.395	0.395	0.389	0.358	0.329	0.275	0.147	0
	0.05	0	0.100	0.237	0.264	0.264	0.262	0.229	0.173	0.101	0.029	0
	0.10	0	0.074	0.125	0.108	0.108	0.086	0.060	0.032	0.008	−0.005	0
	0.15	0	0.040	0.043	0.024	0.024	0.005	−0.010	−0.020	−0.022	−0.016	0
	0.20	0	0.013	−0.002	−0.015	−0.015	−0.027	−0.034	−0.035	−0.030	−0.019	0
	0.25	0	−0.004	−0.021	−0.029	−0.029	−0.035	−0.037	−0.035	−0.028	−0.017	0
	0.30	0	−0.012	−0.027	−0.030	−0.030	−0.033	−0.033	−0.029	−0.023	−0.013	0
	0.40	0	−0.013	−0.021	−0.021	−0.021	−0.021	−0.019	−0.016	−0.012	−0.007	0
	0.50	0	−0.008	−0.012	−0.011	−0.011	−0.010	−0.009	−0.007	−0.005	−0.003	0
	0.60	0	−0.004	−0.005	−0.005	−0.005	−0.004	−0.003	−0.003	−0.002	−0.001	0
	0.70	0	−0.002	−0.002	−0.001	−0.001	−0.001	−0.001	−0.001	−0.001	0.000	0
	0.80	0	0.000	0.000	0.000	0.000	0.000	0.000	0.000	0.000	0.000	0

$\gamma = 2; \mu = 0{,}75; \delta = 0{,}5$

	λ	Core					Shell					
	p	0	0.3	0.6	0.7	0.75	0.75	0.80	0.85	0.90	0.95	1.00
$\dfrac{\bar{\sigma}_z}{\eta_1}$	0	0	0	0	0	−0.228	0.261	0	0	0	0	0.6
	0.01	−0.021	−0.023	−0.037	−0.076	−0.257	0.188	0.006	−0.007	−0.003	0.037	0.613
	0.02	−0.042	−0.046	−0.074	−0.146	−0.281	0.126	0.019	0.001	0.027	0.152	0.617
	0.05	−0.102	−0.112	−0.174	−0.255	−0.316	0.024	0.049	0.094	0.193	0.372	0.624
	0.10	−0.188	−0.200	−0.255	−0.282	−0.290	0.051	0.130	0.221	0.332	0.462	0.610
	0.15	−0.245	−0.253	−0.267	−0.259	−0.248	0.128	0.197	0.275	0.360	0.452	0.548
	0.20	−0.273	−0.273	−0.255	−0.234	−0.218	0.186	0.237	0.293	0.351	0.412	0.472
	0.25	−0.279	−0.272	−0.238	−0.215	−0.200	0.221	0.256	0.293	0.331	0.369	0.405
	0.30	−0.270	−0.260	−0.223	−0.203	−0.191	0.240	0.263	0.286	0.309	0.332	0.352
	0.40	−0.240	−0.230	−0.203	−0.191	−0.186	0.251	0.260	0.268	0.275	0.282	0.287
	0.50	−0.214	−0.209	−0.194	−0.188	−0.186	0.251	0.253	0.255	0.257	0.257	0.258
	0.60	−0.200	−0.197	−0.191	−0.188	−0.187	0.247	0.248	0.248	0.248	0.247	0.247
	0.70	−0.193	−0.192	−0.189	−0.189	−0.188	0.246	0.245	0.245	0.244	0.244	0.243
	0.80	−0.189	−0.189	−0.189	−0.189	−0.189	0.244	0.244	0.244	0.244	0.244	0.244
$\dfrac{\bar{\sigma}_\theta}{\eta_1}$	0	−0.6	−0.6	−0.6	−0.6	−0.635	−0.554	−0.6	−0.6	−0.6	−0.6	−0.4
	0.01	−0.573	−0.572	−0.570	−0.574	−0.592	−0.483	−0.502	−0.497	−0.479	−0.436	−0.287
	0.02	−0.547	−0.545	−0.539	−0.541	−0.548	−0.409	−0.412	−0.399	−0.369	−0.312	−0.209
	0.05	−0.469	−0.464	−0.446	−0.434	−0.424	−0.195	−0.181	−0.158	−0.126	−0.085	−0.038
	0.10	−0.351	−0.342	−0.309	−0.289	−0.276	0.078	0.084	0.094	0.108	0.125	0.143
	0.15	−0.256	−0.247	−0.215	−0.198	−0.189	0.247	0.239	0.235	0.234	0.236	0.240
	0.20	−0.188	−0.181	−0.157	−0.146	−0.140	0.342	0.324	0.310	0.300	0.293	0.288
	0.25	−0.145	−0.140	−0.125	−0.119	−0.116	0.391	0.366	0.347	0.331	0.318	0.308
	0.30	−0.120	−0.117	−0.109	−0.105	−0.104	0.413	0.385	0.362	0.343	0.327	0.313
	0.40	−0.102	−0.102	−0.100	−0.101	−0.101	0.421	0.390	0.364	0.342	0.323	0.308
	0.50	−0.103	−0.103	−0.104	−0.104	−0.104	0.414	0.382	0.356	0.334	0.315	0.299
	0.60	−0.106	−0.106	−0.108	−0.108	−0.108	0.407	0.375	0.349	0.327	0.308	0.292
	0.70	−0.109	−0.109	−0.110	−0.110	−0.110	0.402	0.371	0.345	0.323	0.305	0.289
	0.80	−0.112	−0.112	−0.112	−0.112	−0.112	0.400	0.368	0.343	0.321	0.303	0.288
$\dfrac{\bar{\sigma}_r}{\eta_1}$	0	−0.6	−0.6	−0.6	−0.6	−0.584	−0.584	−0.6	−0.6	−0.6	−0.6	0
	0.01	−0.573	−0.571	−0.561	−0.559	−0.505	−0.505	−0.453	−0.429	−0.364	−0.205	0
	0.02	−0.547	−0.542	−0.519	−0.494	−0.432	−0.432	−0.354	−0.293	−0.198	−0.060	0
	0.05	−0.469	−0.456	−0.392	−0.324	−0.269	−0.269	−0.189	−0.110	−0.044	−0.006	0
	0.10	−0.351	−0.329	−0.239	−0.184	−0.155	−0.155	−0.104	−0.062	−0.030	−0.010	0
	0.15	−0.256	−0.234	−0.164	−0.134	−0.121	−0.121	−0.084	−0.054	−0.031	−0.013	0
	0.20	−0.188	−0.172	−0.130	−0.117	−0.111	−0.111	−0.079	−0.053	−0.032	−0.015	0
	0.25	−0.145	−0.135	−0.116	−0.111	−0.110	−0.110	−0.079	−0.054	−0.033	−0.016	0
	0.30	−0.120	−0.116	−0.110	−0.110	−0.110	−0.110	−0.080	−0.055	−0.034	−0.016	0
	0.40	−0.102	−0.103	−0.108	−0.110	−0.111	−0.111	−0.081	−0.056	−0.034	−0.016	0
	0.50	−0.103	−0.105	−0.109	−0.111	−0.112	−0.112	−0.081	−0.056	−0.034	−0.016	0
	0.60	−0.106	−0.108	−0.110	−0.111	−0.112	−0.112	−0.081	−0.055	−0.034	−0.016	0
	0.70	−0.109	−0.110	−0.111	−0.112	−0.112	−0.112	−0.081	−0.055	−0.034	−0.016	0
	0.80	−0.112	−0.112	−0.112	−0.112	−0.112	−0.112	−0.081	−0.055	−0.034	−0.016	0
$\dfrac{\bar{\tau}_{rz}}{\eta_1}$	0	0	0.105	0.257	0.368	$+\infty$	$+\infty$	0.425	0.412	0.406	0.399	0
	0.01	0	0.105	0.256	0.371	0.477	0.477	0.426	0.397	0.370	0.287	0
	0.02	0	0.104	0.253	0.358	0.411	0.411	0.402	0.357	0.293	0.157	0
	0.05	0	0.095	0.219	0.262	0.261	0.261	0.245	0.192	0.116	0.037	0
	0.10	0	0.069	0.121	0.112	0.096	0.096	0.066	0.036	0.011	−0.003	0
	0.15	0	0.037	0.042	0.026	0.015	0.015	−0.005	−0.018	−0.023	−0.017	0
	0.20	0	0.011	−0.002	−0.014	−0.020	−0.020	−0.030	−0.034	−0.030	−0.019	0
	0.25	0	−0.005	−0.021	−0.028	−0.031	−0.031	−0.035	−0.034	−0.028	−0.017	0
	0.30	0	−0.013	−0.028	−0.030	−0.031	−0.031	−0.031	−0.029	−0.022	−0.013	0
	0.40	0	−0.014	−0.022	−0.021	−0.021	−0.021	−0.019	−0.016	−0.012	−0.006	0
	0.50	0	−0.009	−0.012	−0.011	−0.010	−0.010	−0.009	−0.011	−0.005	−0.003	0
	0.60	0	−0.004	−0.006	−0.005	−0.005	−0.005	−0.004	−0.005	−0.002	−0.001	0
	0.70	0	−0.002	−0.002	−0.002	−0.002	−0.002	−0.001	−0.002	−0.001	0.000	0
	0.80	0	0.000	0.000	0.000	0.000	0.000	0.000	0.000	0.000	0.000	0

	λ \\ p	Core						Shell				
		0	0.3	0.6	0.7	0.75	0.80	0.80	0.85	0.90	0.95	1.00
$\dfrac{\bar{\sigma}_z}{\eta_1}$	0	0	0	0	0	0	−0.228	0.261	0	0	0	0.6
	0.01	−0.020	−0.022	−0.032	−0.048	−0.077	−0.259	0.181	0.010	0.004	0.044	0.622
	0.02	−0.041	−0.044	−0.064	−0.095	−0.146	−0.282	0.118	0.032	0.042	0.167	0.635
	0.05	−0.099	−0.107	−0.149	−0.199	−0.243	−0.297	0.050	0.117	0.222	0.408	0.672
	0.10	−0.181	−0.190	−0.226	−0.243	−0.247	−0.243	0.137	0.243	0.366	0.510	0.672
	0.15	−0.234	−0.238	−0.240	−0.228	−0.215	−0.196	0.228	0.310	0.402	0.500	0.603
	0.20	−0.258	−0.255	−0.229	−0.206	−0.188	−0.169	0.280	0.337	0.398	0.461	0.524
	0.25	−0.260	−0.251	−0.213	−0.188	−0.173	−0.157	0.306	0.342	0.381	0.419	0.456
	0.30	−0.249	−0.238	−0.198	−0.177	−0.165	−0.152	0.314	0.337	0.360	0.383	0.403
	0.40	−0.216	−0.207	−0.179	−0.166	−0.161	−0.154	0.310	0.319	0.326	0.333	0.338
	0.50	−0.190	−0.184	−0.169	−0.163	−0.161	−0.158	0.303	0.305	0.306	0.307	0.308
	0.60	−0.175	−0.172	−0.166	−0.163	−0.162	−0.161	0.296	0.297	0.297	0.296	0.296
	0.70	−0.168	−0.167	−0.164	−0.163	−0.163	−0.163	0.294	0.293	0.293	0.292	0.292
	0.80	−0.164	−0.164	−0.164	−0.164	−0.164	−0.164	0.291	0.291	0.291	0.291	0.291
$\dfrac{\bar{\sigma}_\theta}{\eta_1}$	0	−0.6	−0.6	−0.6	−0.6	−0.6	−0.635	−0.554	−0.6	−0.6	−0.6	−0.4
	0.01	−0.572	−0.571	−0.567	−0.566	−0.568	−0.585	−0.473	−0.487	−0.473	−0.430	−0.281
	0.02	−0.544	−0.542	−0.533	−0.530	−0.531	−0.534	−0.389	−0.384	−0.357	−0.301	−0.198
	0.05	−0.462	−0.456	−0.435	−0.422	−0.413	−0.400	−0.155	−0.132	−0.100	−0.058	−0.010
	0.10	−0.338	−0.329	−0.295	−0.275	−0.263	−0.248	0.128	0.137	0.151	0.167	0.186
	0.15	−0.241	−0.230	−0.199	−0.182	−0.173	−0.163	0.294	0.287	0.285	0.285	0.287
	0.20	−0.171	−0.164	−0.140	−0.129	−0.123	−0.117	0.384	0.366	0.353	0.343	0.336
	0.25	−0.127	−0.122	−0.107	−0.101	−0.098	−0.095	0.428	0.405	0.385	0.370	0.356
	0.30	−0.102	−0.099	−0.091	−0.088	−0.087	−0.085	0.448	0.420	0.398	0.379	0.362
	0.40	−0.085	−0.084	−0.083	−0.083	−0.083	−0.083	0.453	0.423	0.397	0.376	0.357
	0.50	−0.085	−0.085	−0.086	−0.086	−0.087	−0.087	0.445	0.414	0.389	0.367	0.348
	0.60	−0.089	−0.089	−0.090	−0.090	−0.090	−0.091	0.438	0.407	0.382	0.360	0.342
	0.70	−0.092	−0.092	−0.092	−0.092	−0.093	−0.093	0.433	0.403	0.378	0.356	0.338
	0.80	−0.094	−0.094	−0.094	−0.094	−0.094	−0.094	0.430	0.400	0.376	0.355	0.336
$\dfrac{\bar{\sigma}_r}{\eta_1}$	0	−0.6	−0.6	−0.6	−0.6	−0.6	−0.584	−0.584	−0.6	−0.6	−0.6	0
	0.01	−0.572	−0.569	−0.556	−0.549	−0.545	−0.486	−0.486	−0.418	−0.361	−0.204	0
	0.02	−0.544	−0.538	−0.513	−0.493	−0.469	−0.397	−0.397	−0.297	−0.198	−0.060	0
	0.05	−0.462	−0.448	−0.385	−0.329	−0.282	−0.220	−0.220	−0.133	−0.056	−0.009	0
	0.10	−0.338	−0.316	−0.230	−0.178	−0.148	−0.120	−0.120	−0.074	−0.038	−0.013	0
	0.15	−0.241	−0.219	−0.151	−0.122	−0.108	−0.097	−0.097	−0.062	−0.036	−0.016	0
	0.20	−0.171	−0.156	−0.115	−0.102	−0.097	−0.093	−0.093	−0.062	−0.037	−0.017	0
	0.25	−0.127	−0.118	−0.100	−0.095	−0.094	−0.093	−0.093	−0.063	−0.039	−0.018	0
	0.30	−0.102	−0.098	−0.093	−0.093	−0.093	−0.094	−0.094	−0.064	−0.040	−0.019	0
	0.40	−0.085	−0.086	−0.090	−0.092	−0.094	−0.095	−0.095	−0.065	−0.040	−0.019	0
	0.50	−0.085	−0.087	−0.091	−0.093	−0.094	−0.095	−0.095	−0.065	−0.040	−0.019	0
	0.60	−0.089	−0.090	−0.093	−0.094	−0.094	−0.095	−0.094	−0.065	−0.040	−0.019	0
	0.70	−0.092	−0.092	−0.094	−0.094	−0.094	−0.095	−0.095	−0.065	−0.040	−0.019	0
	0.80	−0.094	−0.094	−0.094	−0.094	−0.094	−0.094	−0.094	−0.065	−0.040	−0.019	0
$\dfrac{\bar{\tau}_{rz}}{\eta_1}$	0	0	0.102	0.239	0.318	0.387	$+\infty$	$+\infty$	0.451	0.430	0.411	0
	0.01	0	0.101	0.238	0.317	0.388	0.500	0.500	0.444	0.396	0.300	0
	0.02	0	0.100	0.233	0.310	0.371	0.421	0.421	0.402	0.321	0.171	0
	0.05	0	0.091	0.202	0.245	0.252	0.248	0.248	0.208	0.133	0.048	0
	0.10	0	0.064	0.113	0.108	0.097	0.077	0.077	0.041	0.013	−0.003	0
	0.15	0	0.033	0.038	0.025	0.015	0.003	0.003	−0.015	−0.022	−0.018	0
	0.20	0	0.008	−0.005	−0.015	−0.020	−0.025	−0.025	−0.031	−0.030	−0.019	0
	0.25	0	−0.008	−0.024	−0.029	−0.031	−0.032	−0.032	−0.032	−0.027	−0.016	0
	0.30	0	−0.014	−0.029	−0.031	−0.031	−0.030	−0.030	−0.028	−0.022	−0.013	0
	0.40	0	−0.014	−0.022	−0.022	−0.021	−0.020	−0.020	−0.016	−0.012	−0.006	0
	0.50	0	−0.009	−0.013	−0.012	−0.011	−0.010	−0.010	−0.008	−0.006	−0.003	0
	0.60	0	−0.004	−0.006	−0.005	−0.004	−0.004	−0.004	−0.003	−0.002	−0.001	0
	0.70	0	−0.002	−0.002	−0.002	−0.002	−0.002	−0.002	−0.001	−0.001	0.000	0
	0.80	0	0.000	0.000	0.000	0.000	0.000	0.000	0.000	0.000	0.000	0

$\gamma = \cdot 2;\ \mu = 0{,}85;\ \delta = 0{,}5$

	λ	Core							Shell			
	p	0	0.3	0.6	0.7	0.75	0.80	0.85	0.85	0.90	0.95	1.00
$\bar{\sigma}_z$ η_1	0	0	0	0	0	0	0	−0.228	0.261	0	0	0.6
	0.01	−0.020	−0.021	−0.029	−0.037	−0.048	−0.077	−0.258	0.169	0.022	0.056	0.636
	0.02	−0.040	−0.042	−0.057	−0.074	−0.094	−0.144	−0.278	0.111	0.076	0.193	0.666
	0.05	−0.096	−0.102	−0.131	−0.160	−0.186	−0.221	−0.258	0.117	0.254	0.458	0.749
	0.10	−0.174	−0.180	−0.200	−0.205	−0.203	−0.194	−0.173	0.272	0.409	0.570	0.750
	0.15	−0.222	−0.223	−0.211	−0.192	−0.176	−0.154	−0.125	0.364	0.459	0.563	0.672
	0.20	−0.242	−0.235	−0.199	−0.172	−0.154	−0.132	−0.107	0.400	0.461	0.526	0.590
	0.25	−0.240	−0.228	−0.183	−0.156	−0.140	−0.123	−0.104	0.407	0.445	0.484	0.521
	0.30	−0.226	−0.213	−0.168	−0.146	−0.134	−0.120	−0.107	0.401	0.424	0.447	0.468
	0.40	−0.189	−0.179	−0.149	−0.136	−0.130	−0.123	−0.117	0.380	0.388	0.395	0.400
	0.50	−0.161	−0.155	−0.139	−0.133	−0.131	−0.128	−0.126	0.364	0.366	0.367	0.367
	0.60	−0.145	−0.142	−0.135	−0.133	−0.132	−0.131	−0.130	0.355	0.355	0.354	0.353
	0.70	−0.138	−0.136	−0.134	−0.133	−0.133	−0.133	−0.132	0.350	0.350	0.349	0.349
	0.80	−0.134	−0.134	−0.134	−0.134	−0.134	−0.134	−0.134	0.348	0.348	0.348	0.348
$\bar{\sigma}_\theta$ η_1	0	−0.6	−0.6	−0.6	−0.6	−0.6	−0.6	−0.635	−0.554	−0.6	−0.6	−0.4
	0.01	−0.570	−0.569	−0.564	−0.561	−0.559	−0.561	−0.575	−0.459	−0.462	−0.422	−0.273
	0.02	−0.540	−0.538	−0.527	−0.521	−0.518	−0.516	−0.515	−0.362	−0.340	−0.285	−0.182
	0.05	−0.453	−0.447	−0.423	−0.408	−0.398	−0.385	−0.367	−0.101	−0.066	−0.022	0.028
	0.10	−0.324	−0.314	−0.278	−0.257	−0.244	−0.230	−0.213	0.192	0.203	0.220	0.238
	0.15	−0.222	−0.212	−0.179	−0.162	−0.153	−0.143	−0.132	0.350	0.344	0.342	0.342
	0.20	−0.151	−0.143	−0.119	−0.109	−0.103	−0.097	−0.091	0.433	0.416	0.403	0.392
	0.25	−0.106	−0.101	−0.087	−0.081	−0.077	−0.074	−0.071	0.472	0.449	0.430	0.413
	0.30	−0.081	−0.078	−0.070	−0.068	−0.066	−0.065	−0.063	0.488	0.461	0.439	0.419
	0.40	−0.064	−0.063	−0.063	−0.063	−0.063	−0.063	−0.063	0.490	0.460	0.435	0.414
	0.50	−0.064	−0.065	−0.066	−0.066	−0.066	−0.067	−0.067	0.481	0.451	0.426	0.404
	0.60	−0.069	−0.069	−0.070	−0.070	−0.070	−0.071	−0.071	0.473	0.443	0.418	0.397
	0.70	−0.072	−0.072	−0.073	−0.073	−0.073	−0.073	−0.073	0.469	0.439	0.414	0.393
	0.80	−0.075	−0.075	−0.075	−0.075	−0.075	−0.075	−0.075	0.465	0.436	0.411	0.390
$\bar{\sigma}_r$ η_1	0	−0.6	−0.6	−0.6	−0.6	−0.6	−0.6	−0.584	−0.584	−0.6	−0.6	0
	0.01	−0.570	−0.567	−0.553	−0.542	−0.535	−0.525	−0.456	−0.456	−0.353	−0.202	0
	0.02	−0.540	−0.534	−0.506	−0.484	−0.466	−0.433	−0.345	−0.345	−0.206	−0.061	0
	0.05	−0.453	−0.438	−0.374	−0.324	−0.284	−0.229	−0.161	−0.161	−0.075	−0.017	0
	0.10	−0.324	−0.300	−0.216	−0.167	−0.139	−0.110	−0.084	−0.084	−0.044	−0.016	0
	0.15	−0.222	−0.201	−0.135	−0.107	−0.094	−0.082	−0.073	−0.073	−0.042	−0.018	0
	0.20	−0.151	−0.136	−0.097	−0.085	−0.080	−0.076	−0.073	−0.073	−0.044	−0.020	0
	0.25	−0.106	−0.098	−0.081	−0.077	−0.076	−0.075	−0.075	−0.075	−0.046	−0.022	0
	0.30	−0.081	−0.078	−0.074	−0.074	−0.074	−0.075	−0.076	−0.076	−0.047	−0.022	0
	0.40	−0.064	−0.065	−0.070	−0.073	−0.074	−0.075	−0.076	−0.076	−0.047	−0.022	0
	0.50	−0.064	−0.066	−0.071	−0.073	−0.074	−0.075	−0.076	−0.076	−0.046	−0.022	0
	0.60	−0.069	−0.070	−0.073	−0.074	−0.075	−0.075	−0.075	−0.075	−0.046	−0.021	0
	0.70	−0.072	−0.072	−0.074	−0.075	−0.075	−0.075	−0.075	−0.075	−0.046	−0.021	0
	0.80	−0.075	−0.075	−0.075	−0.075	−0.075	−0.075	−0.075	−0.075	−0.046	−0.021	0
$\bar{\tau}_{rz}$ η_1	0	0	0.099	0.225	0.289	0.334	0.404	$+\infty$	$+\infty$	0.472	0.433	0
	0.01	0	0.098	0.224	0.287	0.332	0.402	0.511	0.511	0.445	0.324	0
	0.02	0	0.097	0.219	0.279	0.321	0.378	0.420	0.420	0.367	0.196	0
	0.05	0	0.087	0.186	0.223	0.237	0.240	0.217	0.217	0.146	0.058	0
	0.10	0	0.059	0.102	0.099	0.090	0.074	0.052	0.052	0.015	−0.005	0
	0.15	0	0.028	0.031	0.019	0.010	0.000	−0.009	−0.009	−0.021	−0.019	0
	0.20	0	0.004	−0.010	−0.019	−0.023	−0.027	−0.029	−0.029	−0.029	−0.019	0
	0.25	0	−0.011	−0.027	−0.032	−0.033	−0.034	−0.032	−0.032	−0.027	−0.017	0
	0.30	0	−0.017	−0.031	−0.033	−0.033	−0.031	−0.029	−0.029	−0.023	−0.013	0
	0.40	0	−0.012	−0.024	−0.023	−0.022	−0.020	−0.018	−0.018	−0.013	−0.007	0
	0.50	0	−0.007	−0.013	−0.012	−0.011	−0.010	−0.009	−0.009	−0.006	−0.003	0
	0.60	0	−0.003	−0.006	−0.006	−0.005	−0.005	−0.004	−0.004	−0.003	−0.001	0
	0.70	0	−0.001	−0.002	−0.002	−0.002	−0.002	−0.001	−0.001	−0.001	0.000	0
	0.80	0	0.000	0.000	0.000	0.000	0.000	0.000	0.000	0.000	0.000	0

$\gamma = 2; \; \mu = 0.9; \delta = 0.5$

	λ	Core								Shell		
	ρ	0	0.3	0.6	0.7	0.75	0.80	0.85	0.90	0.90	0.95	1.00
$\dfrac{\bar{\sigma}_z}{\eta_1}$	0	0	0	0	0	0	0	0	−0.228	0.261	0	0.6
	0.01	−0.019	−0.021	−0.026	−0.032	−0.036	−0.046	−0.075	−0.261	0.162	0.082	0.667
	0.02	−0.039	−0.041	−0.052	−0.062	−0.071	−0.089	−0.136	−0.260	0.129	0.243	0.734
	0.05	−0.094	−0.098	−0.117	−0.131	−0.142	−0.157	−0.173	−0.172	0.274	0.527	0.869
	0.10	−0.167	−0.171	−0.174	−0.167	−0.157	−0.139	−0.110	−0.064	0.481	0.655	0.850
	0.15	−0.210	−0.207	−0.180	−0.152	−0.131	−0.105	−0.071	−0.031	0.548	0.654	0.766
	0.20	−0.224	−0.213	−0.165	−0.132	−0.111	−0.087	−0.059	−0.029	0.551	0.616	0.682
	0.25	−0.216	−0.202	−0.148	−0.117	−0.100	−0.081	−0.061	−0.040	0.531	0.571	0.610
	0.30	−0.198	−0.183	−0.133	−0.108	−0.094	−0.081	−0.066	−0.053	0.506	0.530	0.551
	0.40	−0.156	−0.145	−0.112	−0.099	−0.092	−0.086	−0.079	−0.073	0.464	0.470	0.475
	0.50	−0.126	−0.119	−0.103	−0.097	−0.094	−0.092	−0.089	−0.087	0.436	0.438	0.438
	0.60	−0.109	−0.106	−0.099	−0.097	−0.096	−0.095	−0.094	−0.093	0.424	0.423	0.422
	0.70	−0.101	−0.100	−0.098	−0.097	−0.097	−0.096	−0.096	−0.096	0.418	0.417	0.417
	0.80	−0.098	−0.098	−0.098	−0.098	−0.098	−0.098	−0.098	−0.098	0.415	0.415	0.415
$\dfrac{\bar{\sigma}_\theta}{\eta_1}$	0		−0.6	−0.6	−0.6	−0.6	−0.6	−0.6	−0.635	−0.554	−0.6	−0.4
	0.01	−0.568	−0.566	−0.560	−0.556	−0.554	−0.551	−0.549	−0.557	−0.435	−0.407	−0.259
	0.02	−0.536	−0.533	−0.521	−0.513	−0.508	−0.501	−0.494	−0.484	−0.317	−0.260	−0.155
	0.05	−0.443	−0.436	−0.409	−0.392	−0.380	−0.365	−0.346	−0.320	−0.022	0.027	0.085
	0.10	−0.306	−0.295	−0.257	−0.234	−0.221	−0.206	−0.189	−0.170	0.271	0.285	0.303
	0.15	−0.199	−0.189	−0.155	−0.137	−0.128	−0.117	−0.107	−0.096	0.418	0.412	0.409
	0.20	−0.126	−0.118	−0.094	−0.083	−0.077	−0.071	−0.066	−0.060	0.491	0.474	0.460
	0.25	−0.081	−0.075	−0.061	−0.055	−0.052	−0.049	−0.046	−0.043	0.524	0.501	0.481
	0.30	−0.055	−0.053	−0.045	−0.043	−0.041	−0.040	−0.039	−0.037	0.536	0.509	0.486
	0.40	−0.039	−0.039	−0.039	−0.039	−0.039	−0.039	−0.039	−0.039	0.533	0.503	0.478
	0.50	−0.041	−0.042	−0.043	−0.043	−0.044	−0.044	−0.044	−0.045	0.522	0.492	0.467
	0.60	−0.046	−0.046	−0.047	−0.048	−0.048	−0.048	−0.049	−0.049	0.513	0.484	0.459
	0.70	−0.050	−0.050	−0.050	−0.051	−0.051	−0.051	−0.051	−0.051	0.508	0.479	0.455
	0.80	−0.053	−0.053	−0.053	−0.053	−0.053	−0.053	−0.053	−0.053	0.505	0.476	0.452
$\dfrac{\bar{\sigma}_r}{\eta_1}$	0		−0.6	−0.6	−0.6	−0.6	−0.6	−0.6	−0.584	−0.584	−0.6	0
	0.01	−0.568	−0.564	−0.548	−0.536	−0.526	−0.513	−0.492	−0.393	−0.393	−0.198	0
	0.02	−0.536	−0.529	−0.498	−0.473	−0.454	−0.426	−0.376	−0.258	−0.258	−0.073	0
	0.05	−0.443	−0.427	−0.359	−0.309	−0.272	−0.225	−0.163	−0.094	−0.094	−0.026	0
	0.10	−0.306	−0.282	−0.196	−0.149	−0.122	−0.095	−0.071	−0.051	−0.051	−0.019	0
	0.15	−0.199	−0.178	−0.113	−0.087	−0.075	−0.064	−0.056	−0.051	−0.051	−0.022	0
	0.20	−0.126	−0.111	−0.075	−0.064	−0.060	−0.056	−0.054	−0.053	−0.053	−0.024	0
	0.25	−0.081	−0.073	−0.058	−0.055	−0.054	−0.054	−0.054	−0.054	−0.054	−0.025	0
	0.30	−0.055	−0.053	−0.050	−0.051	−0.052	−0.053	−0.053	−0.054	−0.054	−0.026	0
	0.40	−0.039	−0.041	−0.047	−0.050	−0.051	−0.053	−0.053	−0.055	−0.055	−0.025	0
	0.50	−0.041	−0.043	−0.049	−0.051	−0.052	−0.052	−0.053	−0.054	−0.054	−0.025	0
	0.60	−0.046	−0.047	−0.051	−0.052	−0.052	−0.053	−0.053	−0.054	−0.054	−0.025	0
	0.70	−0.050	−0.050	−0.052	−0.052	−0.053	−0.053	−0.053	−0.053	−0.053	−0.025	0
	0.80	−0.053	−0.053	−0.053	−0.053	−0.053	−0.053	−0.053	−0.053	−0.053	−0.024	0
$\dfrac{\bar{\tau}_{rz}}{\eta_1}$	0	0	0.096	0.214	0.268	0.303	0.347	0.416	$+\infty$	$+\infty$	0.476	0
	0.01	0	0.095	0.212	0.265	0.298	0.342	0.409	0.510	0.510	0.372	0
	0.02	0	0.094	0.207	0.256	0.286	0.323	0.371	0.393	0.393	0.237	0
	0.05	0	0.084	0.172	0.200	0.211	0.215	0.203	0.161	0.161	0.063	0
	0.10	0	0.054	0.088	0.083	0.074	0.060	0.042	0.021	0.021	−0.005	0
	0.15	0	0.022	0.019	0.008	0.000	−0.008	−0.015	−0.020	−0.020	−0.019	0
	0.20	0	−0.001	−0.018	−0.026	−0.030	−0.032	−0.032	−0.030	−0.030	−0.020	0
	0.25	0	−0.015	−0.033	−0.037	−0.037	−0.037	−0.034	−0.030	−0.030	−0.018	0
	0.30	0	−0.020	−0.035	−0.036	−0.036	−0.034	−0.030	−0.026	−0.026	−0.015	0
	0.40	0	−0.017	−0.025	−0.024	−0.023	−0.021	−0.018	−0.015	−0.015	−0.008	0
	0.50	0	−0.010	−0.014	−0.013	−0.012	−0.010	−0.009	−0.007	−0.007	−0.004	0
	0.60	0	−0.005	−0.006	−0.006	−0.005	−0.005	−0.004	−0.003	−0.003	−0.002	0
	0.70	0	−0.002	−0.002	−0.002	−0.002	−0.001	−0.001	−0.001	−0.001	0.000	0
	0.80	0	0.000	0.000	0.000	0.000	0.000	0.000	0.000	0.000	0.000	0

γ = 2; μ = 0,95; δ = 0,5

	λ	Core									Shell	
	p	0	0.3	0.6	0.7	0.75	0.80	0.85	0.90	0.95	0.95	1.00
$\bar{\sigma}_s/\eta_1$	0	0	0	0	0	0	0	0	0	-0.228	0.261	0.6
	0.01	-0.019	-0.020	-0.024	-0.027	-0.029	-0.034	-0.042	-0.066	-0.237	0.180	0.770
	0.02	-0.038	-0.040	-0.047	-0.053	-0.056	-0.063	-0.074	-0.103	-0.162	0.302	0.924
	0.05	-0.092	-0.095	-0.105	-0.108	-0.108	-0.104	-0.092	-0.059	0.021	0.650	1.059
	0.10	-0.161	-0.162	-0.148	-0.125	-0.105	-0.075	-0.032	0.027	0.104	0.813	1.017
	0.15	-0.197	-0.190	-0.144	-0.104	-0.076	-0.042	-0.001	0.046	0.099	0.803	0.920
	0.20	-0.203	-0.188	-0.124	-0.082	-0.057	-0.028	0.003	0.037	0.072	0.750	0.818
	0.25	-0.189	-0.170	-0.104	-0.068	-0.047	-0.026	-0.003	0.019	0.042	0.690	0.729
	0.30	-0.165	-0.146	-0.087	-0.059	-0.044	-0.029	-0.014	0.012	0.015	0.636	0.658
	0.40	-0.116	-0.102	-0.067	-0.053	-0.046	-0.039	-0.033	-0.027	-0.022	0.561	0.566
	0.50	-0.082	-0.075	-0.058	-0.052	-0.049	-0.047	-0.045	-0.043	-0.042	0.522	0.522
	0.60	-0.065	-0.062	-0.055	-0.053	-0.052	-0.051	-0.050	-0.050	-0.050	0.505	0.504
	0.70	-0.057	-0.056	-0.054	-0.053	-0.053	-0.053	-0.053	-0.053	-0.053	0.499	0.498
	0.80	-0.054	-0.054	-0.054	-0.054	-0.054	-0.054	-0.054	-0.054	-0.054	0.498	0.497
$\bar{\sigma}_\theta/\eta_1$	0	-0.6	-0.6	-0.6	-0.6	-0.6	-0.6	-0.6	-0.6	-0.635	-0.554	-0.4
	0.01	-0.565	-0.563	-0.556	-0.551	-0.548	-0.543	-0.536	-0.527	-0.517	-0.383	-0.227
	0.02	-0.531	-0.527	-0.513	-0.503	-0.496	-0.486	-0.475	-0.456	-0.424	-0.225	-0.094
	0.05	-0.430	-0.422	-0.391	-0.370	-0.356	-0.339	-0.317	-0.288	-0.252	0.102	0.166
	0.10	-0.282	-0.270	-0.227	-0.204	-0.189	-0.173	-0.155	-0.135	-0.115	0.376	0.390
	0.15	-0.170	-0.158	-0.122	-0.104	-0.094	-0.083	-0.072	-0.061	-0.050	0.505	0.498
	0.20	-0.094	-0.085	-0.060	-0.049	-0.044	-0.038	-0.032	-0.026	-0.021	0.564	0.547
	0.25	-0.048	-0.043	-0.029	-0.023	-0.020	-0.017	-0.014	-0.012	-0.009	0.588	0.564
	0.30	-0.024	-0.021	-0.014	-0.012	-0.011	-0.010	-0.009	-0.008	-0.007	0.592	0.565
	0.40	-0.011	-0.011	-0.011	-0.011	-0.011	-0.011	-0.012	-0.012	-0.012	0.581	0.552
	0.50	-0.015	-0.015	-0.017	-0.017	-0.018	-0.018	-0.019	-0.019	-0.019	0.567	0.538
	0.60	-0.021	-0.021	-0.022	-0.023	-0.023	-0.023	-0.024	-0.024	-0.024	0.557	0.529
	0.70	-0.025	-0.025	-0.026	-0.026	-0.026	-0.026	-0.026	-0.026	-0.027	0.552	0.524
	0.80	-0.028	-0.028	-0.028	-0.028	-0.028	-0.028	-0.028	-0.028	-0.028	0.550	0.521
$\bar{\sigma}_r/\eta_1$	0	-0.6	-0.6	-0.6	-0.6	-0.6	-0.6	-0.6	-0.6	-0.584	-0.584	0
	0.01	-0.565	-0.561	-0.543	-0.528	-0.517	-0.500	-0.475	-0.427	-0.256	-0.257	0
	0.02	-0.531	-0.523	-0.487	-0.459	-0.438	-0.407	-0.360	-0.278	-0.117	-0.117	0
	0.05	-0.430	-0.412	-0.337	-0.283	-0.247	-0.201	-0.146	-0.084	-0.030	-0.030	0
	0.10	-0.282	-0.257	-0.168	-0.121	-0.096	-0.072	-0.050	-0.034	-0.025	-0.025	0
	0.15	-0.170	-0.147	-0.084	-0.059	-0.049	-0.040	-0.033	-0.029	-0.027	-0.027	0
	0.20	-0.094	-0.079	-0.046	-0.036	-0.033	-0.031	-0.029	-0.029	-0.029	-0.029	0
	0.25	-0.048	-0.041	-0.029	-0.028	-0.028	-0.028	-0.028	-0.029	-0.029	-0.029	0
	0.30	-0.024	-0.022	-0.023	-0.025	-0.026	-0.027	-0.028	-0.029	-0.030	-0.030	0
	0.40	-0.011	-0.013	-0.021	-0.024	-0.026	-0.027	-0.028	-0.029	-0.029	-0.029	0
	0.50	-0.015	-0.017	-0.023	-0.026	-0.027	-0.027	-0.028	-0.029	-0.029	-0.029	0
	0.60	-0.021	-0.022	-0.026	-0.027	-0.027	-0.028	-0.028	-0.028	-0.028	-0.028	0
	0.70	-0.025	-0.025	-0.027	-0.028	-0.028	-0.028	-0.028	-0.028	-0.028	-0.028	0
	0.80	-0.028	-0.028	-0.028	-0.028	-0.028	-0.028	-0.028	-0.028	-0.028	-0.028	0
$\bar{\tau}_{rs}/\eta_1$	0	0	0.094	0.205	0.251	0.278	0.311	0.351	0.415	$+\infty$	$+\infty$	0
	0.01	0	0.093	0.202	0.248	0.274	0.303	0.340	0.392	0.448	0.448	0
	0.02	0	0.091	0.196	0.237	0.259	0.283	0.307	0.326	0.274	0.274	0
	0.05	0	0.080	0.158	0.177	0.181	0.178	0.163	0.128	0.071	0.071	0
	0.10	0	0.048	0.070	0.061	0.051	0.037	0.020	0.004	-0.006	-0.006	0
	0.15	0	0.015	0.004	-0.009	-0.016	-0.023	-0.027	-0.028	-0.024	-0.024	0
	0.20	0	-0.009	-0.030	-0.038	-0.040	-0.041	-0.039	-0.035	-0.026	-0.026	0
	0.25	0	-0.021	-0.042	-0.045	-0.044	-0.042	-0.038	-0.032	-0.023	-0.023	0
	0.30	0	-0.024	-0.041	-0.041	-0.039	-0.036	-0.032	-0.026	-0.018	-0.018	0
	0.40	0	-0.019	-0.027	-0.026	-0.024	-0.021	-0.018	-0.014	-0.010	-0.010	0
	0.50	0	-0.011	-0.014	-0.013	-0.012	-0.010	-0.009	-0.007	-0.004	-0.004	0
	0.60	0	-0.005	-0.006	-0.005	-0.005	-0.004	-0.003	-0.003	-0.002	-0.002	0
	0.70	0	-0.002	-0.002	-0.002	-0.002	-0.001	-0.001	-0.001	0.000	01000	0
	0.80	0	0.000	0.000	0.000	0.000	0.000	0.000	0.000	0.000	0.000	0

	λ	Core				Shell						
	p	0	0.3	0.6	0.7	0.7	0.75	0.80	0.85	0.90	0.95	1.00
$\frac{\bar{\sigma}_z}{\eta_1}$	0	0	0	0	0.15	−0.15	0	0	0	0	0	0.6
	0.01	−0.007	−0.006	0.012	0.132	−0.167	−0.074	−0.053	−0.046	−0.039	−0.002	0.570
	0.02	−0.014	−0.012	0.021	0.115	−0.187	−0.125	−0.095	−0.077	−0.046	0.074	0.530
	0.05	−0.033	−0.027	0.024	0.081	−0.218	−0.182	−0.139	−0.080	0.019	0.182	0.406
	0.10	−0.049	−0.037	0.036	0.096	−0.204	−0.161	−0.107	−0.038	0.044	0.138	0.234
	0.15	−0.042	−0.023	0.067	0.128	−0.172	−0.133	−0.089	−0.041	0.011	0.062	0.110
	0.20	−0.017	0.007	0.098	0.148	−0.152	−0.123	−0.092	−0.061	−0.031	−0.002	0.023
	0.25	0.017	0.041	0.120	0.157	−0.143	−0.123	−0.103	−0.084	−0.066	−0.050	−0.038
	0.30	0.052	0.073	0.134	0.160	−0.140	−0.127	−0.115	−0.103	−0.093	−0.085	−0.079
	0.40	0.106	0.118	0.148	0.159	−0.141	−0.137	−0.132	−0.129	−0.126	−0.125	−0.125
	0.50	0.135	0.141	0.151	0.156	−0.144	−0.143	−0.142	−0.141	−0.141	−0.141	−0.142
	0.60	0.148	0.150	0.153	0.154	−0.146	−0.146	−0.146	−0.146	−0.146	−0.147	−0.147
	0.70	0.152	0.153	0.153	0.153	−0.147	−0.147	−0.147	−0.147	−0.147	−0.148	−0.148
	0.80	0.153	0.153	0.153	0.153	−0.147	−0.147	−0.147	−0.147	−0.147	−0.148	−0.147
$\frac{\bar{\sigma}_\theta}{\eta_1}$	0	−0.3	−0.3	−0.3	−0.3	−0.6	−0.6	−0.6	−0.6	−0.6	−0.6	−0.4
	0.01	−0.273	−0.272	−0.266	−0.267	−0.567	−0.555	−0.544	−0.532	−0.512	−0.467	−0.318
	0.02	−0.246	−0.243	−0.233	−0.229	−0.529	−0.510	−0.491	−0.468	−0.434	−0.374	−0.269
	0.05	−0.169	−0.162	−0.137	−0.121	−0.421	−0.390	−0.358	−0.323	−0.283	−0.236	−0.187
	0.10	−0.058	−0.048	−0.013	0.007	−0.293	−0.262	−0.232	−0.204	−0.178	−0.152	−0.129
	0.15	0.022	0.030	0.058	0.071	−0.229	−0.202	−0.179	−0.159	−0.141	−0.125	−0.111
	0.20	0.070	0.076	0.092	0.098	−0.202	−0.179	−0.160	−0.144	−0.130	−0.119	−0.109
	0.25	0.094	0.097	0.104	0.106	−0.194	−0.174	−0.157	−0.143	−0.131	−0.122	−0.114
	0.30	0.102	0.103	0.105	0.105	−0.195	−0.176	−0.160	−0.147	−0.137	−0.128	−0.121
	0.40	0.098	0.098	0.096	0.094	−0.206	−0.187	−0.172	−0.159	−0.149	−0.140	−0.133
	0.50	0.089	0.088	0.086	0.085	−0.215	−0.196	−0.180	−0.168	−0.157	−0.149	−0.141
	0.60	0.082	0.082	0.081	0.080	−0.220	−0.201	−0.185	−0.172	−0.162	−0.153	−0.145
	0.70	0.079	0.078	0.078	0.078	−0.222	−0.203	−0.187	−0.174	−0.164	−0.154	−0.147
	0.80	0.077	0.077	0.077	0.077	−0.223	−0.204	−0.188	−0.175	−0.164	−0.155	−0.147
$\frac{\bar{\sigma}_r}{\eta_1}$	0	−0.3	−0.3	−0.3	−0.45	−0.45	−0.6	−0.6	−0.6	−0.6	−0.6	0
	0.01	−0.273	−0.270	−0.268	−0.363	−0.363	−0.451	−0.454	−0.426	−0.361	−0.204	0
	0.02	−0.246	−0.241	−0.232	−0.285	−0.285	−0.329	−0.323	−0.279	−0.191	−0.057	0
	0.05	−0.169	−0.156	−0.117	−0.098	−0.098	−0.098	−0.080	−0.047	−0.013	0.006	0
	0.10	−0.058	−0.040	0.015	0.041	0.041	0.035	0.031	0.026	0.019	0.009	0
	0.15	0.022	0.036	0.069	0.077	0.077	0.060	0.045	0.031	0.019	0.008	0
	0.20	0.070	0.076	0.086	0.085	0.085	0.064	0.046	0.030	0.017	0.007	0
	0.25	0.094	0.094	0.089	0.085	0.085	0.063	0.045	0.030	0.017	0.007	0
	0.30	0.102	0.099	0.089	0.084	0.084	0.062	0.044	0.029	0.017	0.007	0
	0.40	0.098	0.094	0.084	0.081	0.081	0.060	0.043	0.029	0.017	0.008	0
	0.50	0.089	0.086	0.080	0.079	0.079	0.058	0.042	0.028	0.017	0.008	0
	0.60	0.082	0.081	0.078	0.077	0.077	0.058	0.042	0.028	0.017	0.008	0
	0.70	0.079	0.078	0.077	0.077	0.077	0.057	0.041	0.028	0.017	0.008	0
	0.80	0.077	0.077	0.077	0.077	0.077	0.057	0.041	0.028	0.017	0.008	0
$\frac{\bar{\tau}_{rz}}{\eta_1}$	0	0	0.032	0.019	$-\infty$	$-\infty$	0.046	0.142	0.213	0.274	0.332	0
	0.01	0	0.032	0.022	−0.039	−0.039	0.054	0.136	0.197	0.238	0.220	0
	0.02	0	0.030	0.026	0.015	0.015	0.063	0.118	0.156	0.161	0.089	0
	0.05	0	0.023	0.024	0.027	0.027	0.032	0.034	0.023	−0.002	−0.024	0
	0.10	0	0.001	−0.020	−0.035	−0.035	−0.043	−0.051	−0.056	−0.053	−0.036	0
	0.15	0	−0.020	−0.053	−0.064	−0.064	−0.067	−0.066	−0.061	−0.049	−0.029	0
	0.20	0	−0.031	−0.061	−0.065	−0.065	−0.063	−0.059	−0.050	−0.038	−0.021	0
	0.25	0	−0.033	−0.056	−0.055	−0.055	−0.051	−0.046	−0.038	−0.028	−0.015	0
	0.30	0	−0.030	−0.044	−0.042	−0.042	−0.038	−0.033	−0.027	−0.019	−0.010	0
	0.40	0	−0.018	−0.023	−0.020	−0.020	−0.018	−0.015	−0.012	−0.008	−0.004	0
	0.50	0	−0.008	−0.010	−0.008	−0.008	−0.007	−0.006	−0.005	−0.003	−0.002	0
	0.60	0	−0.003	−0.003	−0.003	−0.003	−0.002	−0.002	−0.001	−0.001	0.000	0
	0.70	0	−0.001	−0.001	−0.001	−0.001	0.000	0.000	0.000	0.000	0.000	0
	0.80	0	0.000	0.000	0.000	0.000	0.000	0.000	0.000	0.000	0.000	0

$\gamma = 0{,}5; \mu = 0{,}75; \delta = 1$

	λ \ ρ	Core					Shell					
		0	0.3	0.6	0.7	0.75	0.75	0.80	0.85	0.90	0.95	1.00
$\dfrac{\bar{\sigma}_z}{\eta_1}$	0	0	0	0	0	0.15	−0.15	0	0	0	0	0.6
	0.01	−0.008	−0.007	0.002	0.034	0.130	−0.171	−0.077	−0.056	−0.045	−0.009	0.562
	0.02	−0.016	−0.015	0.002	0.048	0.111	−0.191	−0.128	−0.095	−0.058	0.062	0.515
	0.05	−0.037	−0.034	0.002	0.048	0.085	−0.216	−0.169	−0.105	−0.005	0.157	0.376
	0.10	−0.057	−0.047	0.016	0.073	0.115	−0.185	−0.131	−0.065	0.016	0.106	0.197
	0.15	−0.053	−0.035	0.049	0.107	0.144	−0.156	−0.114	−0.067	−0.018	0.031	0.076
	0.20	−0.030	−0.007	0.079	0.127	0.154	−0.146	−0.117	−0.087	−0.058	−0.031	−0.008
	0.25	0.002	0.025	0.100	0.135	0.154	−0.146	−0.127	−0.109	−0.092	−0.077	−0.066
	0.30	0.035	0.055	0.113	0.138	0.150	−0.150	−0.138	−0.127	−0.118	−0.110	−0.105
	0.40	0.087	0.098	0.127	0.137	0.141	−0.159	−0.155	−0.152	−0.149	−0.148	−0.148
	0.50	0.115	0.120	0.131	0.134	0.135	−0.165	−0.164	−0.163	−0.163	−0.163	−0.164
	0.60	0.127	0.128	0.132	0.132	0.133	−0.167	−0.167	−0.167	−0.168	−0.168	−0.169
	0.70	0.131	0.131	0.132	0.132	0.132	−0.169	−0.169	−0.169	−0.169	−0.169	−0.170
	0.80	0.132	0.132	0.132	0.132	0.132	−0.169	−0.169	−0.169	−0.169	−0.169	−0.169
$\dfrac{\bar{\sigma}_\theta}{\eta_1}$	0	−0.3	−0.3	−0.3	−0.3	−0.3	−0.6	−0.6	−0.6	−0.6	−0.6	−0.4
	0.01	−0.274	−0.272	−0.266	−0.263	−0.263	−0.563	−0.549	−0.535	−0.515	−0.470	−0.320
	0.02	−0.247	−0.244	−0.233	−0.225	−0.221	−0.521	−0.499	−0.475	−0.440	−0.380	−0.276
	0.05	−0.172	−0.165	−0.138	−0.120	−0.108	−0.408	−0.374	−0.338	−0.297	−0.250	−0.201
	0.10	−0.064	−0.054	−0.019	0.001	0.013	−0.287	−0.256	−0.226	−0.199	−0.173	−0.150
	0.15	0.013	0.022	0.049	0.062	0.068	−0.232	−0.206	−0.184	−0.165	−0.148	−0.134
	0.20	0.059	0.065	0.081	0.087	0.090	−0.210	−0.189	−0.171	−0.156	−0.143	−0.132
	0.25	0.082	0.085	0.092	0.094	0.095	−0.205	−0.186	−0.170	−0.157	−0.146	−0.137
	0.30	0.090	0.091	0.093	0.093	0.093	−0.207	−0.189	−0.175	−0.162	−0.152	−0.144
	0.40	0.087	0.086	0.084	0.083	0.082	−0.217	−0.200	−0.186	−0.174	−0.164	−0.156
	0.50	0.077	0.077	0.075	0.074	0.074	−0.226	−0.209	−0.194	−0.182	−0.172	−0.163
	0.60	0.071	0.070	0.069	0.069	0.069	−0.231	−0.213	−0.199	−0.186	−0.176	−0.167
	0.70	0.068	0.067	0.067	0.067	0.067	−0.233	−0.215	−0.200	−0.188	−0.177	−0.168
	0.80	0.066	0.066	0.066	0.066	0.066	−0.234	−0.216	−0.201	−0.189	−0.178	−0.169
$\dfrac{\bar{\sigma}_r}{\eta_1}$	0	−0.3	−0.3	−0.3	−0.3	−0.45	−0.45	−0.6	−0.6	−0.6	−0.6	0
	0.01	−0.274	−0.270	−0.260	−0.274	−0.351	−0.351	−0.430	−0.418	−0.358	−0.203	0
	0.02	−0.247	−0.241	−0.221	−0.230	−0.262	−0.262	−0.292	−0.266	−0.186	−0.056	0
	0.05	−0.172	−0.158	−0.109	−0.084	−0.067	−0.067	−0.057	−0.034	−0.006	0.008	0
	0.10	−0.064	−0.045	0.012	0.038	0.050	0.050	0.041	0.033	0.022	0.010	0
	0.15	0.013	0.027	0.060	0.068	0.070	0.070	0.052	0.036	0.021	0.009	0
	0.20	0.059	0.066	0.075	0.074	0.072	0.072	0.052	0.035	0.020	0.008	0
	0.25	0.082	0.083	0.078	0.074	0.072	0.071	0.051	0.034	0.020	0.008	0
	0.30	0.090	0.088	0.077	0.072	0.070	0.070	0.050	0.033	0.019	0.008	0
	0.40	0.087	0.083	0.073	0.070	0.068	0.068	0.049	0.033	0.019	0.009	0
	0.50	0.077	0.075	0.069	0.068	0.067	0.067	0.048	0.033	0.020	0.009	0
	0.60	0.071	0.070	0.067	0.066	0.066	0.066	0.048	0.032	0.020	0.009	0
	0.70	0.068	0.067	0.066	0.066	0.066	0.066	0.048	0.032	0.020	0.009	0
	0.80	0.066	0.066	0.066	0.066	0.066	0.066	0.047	0.032	0.020	0.009	0
$\dfrac{\bar{\tau}_{rz}}{\eta_1}$	0	0	0.037	0.052	0.002	$-\infty$	$-\infty$	0.074	0.175	0.252	0.320	0
	0.01	0	0.037	0.051	0.013	−0.012	−0.012	0.079	0.161	0.216	0.209	0
	0.02	0	0.036	0.050	0.030	0.032	0.032	0.079	0.127	0.142	0.079	0
	0.05	0	0.027	0.033	0.026	0.024	0.024	0.021	0.011	−0.011	−0.030	0
	0.10	0	0.004	−0.018	−0.035	−0.044	−0.044	−0.053	−0.057	−0.054	−0.036	0
	0.15	0	−0.018	−0.051	−0.062	−0.065	−0.065	−0.064	−0.059	−0.047	−0.028	0
	0.20	0	−0.029	−0.059	−0.062	−0.061	−0.061	−0.056	−0.048	−0.036	−0.020	0
	0.25	0	−0.032	−0.053	−0.052	−0.049	−0.049	−0.044	−0.036	−0.026	−0.014	0
	0.30	0	−0.028	−0.042	−0.040	−0.036	−0.036	−0.032	−0.026	−0.018	−0.009	0
	0.40	0	−0.017	−0.022	−0.019	−0.017	−0.017	−0.015	−0.012	−0.008	−0.004	0
	0.50	0	−0.008	−0.009	−0.008	−0.007	−0.007	−0.006	−0.004	−0.003	−0.001	0
	0.60	0	−0.003	−0.003	−0.003	−0.002	−0.002	−0.002	−0.001	−0.001	0.000	0
	0.70	0	−0.001	−0.001	−0.001	0.000	0.000	0.000	0.000	0.000	0.000	0
	0.80	0	0.000	0.000	0.000	0.000	0.000	0.000	0.000	0.000	0.000	0

	λ	Core						Shell				
	p	0	0.3	0.6	0.7	0.75	0.80	0.80	0.85	0.90	0.95	1.00
$\bar{\sigma}_z / \eta_1$	0	0	0	0	0	0	0.15	-0.15	0	0	0	0.6
	0.01	-0.009	-0.009	-0.004	0.008	0.031	0.127	-0.175	-0.081	-0.057	-0.018	0.551
	0.02	-0.017	-0.017	-0.008	0.013	0.043	0.107	-0.195	-0.129	-0.079	0.045	0.493
	0.05	-0.041	-0.039	-0.016	0.019	0.050	0.097	-0.204	-0.138	-0.036	0.124	0.335
	0.10	-0.064	-0.056	-0.002	0.050	0.090	0.141	-0.159	-0.095	-0.018	0.068	0.153
	0.15	-0.063	-0.047	0.030	0.085	0.119	0.160	-0.141	-0.097	-0.050	-0.005	0.037
	0.20	-0.043	-0.022	0.058	0.104	0.129	0.157	-0.143	-0.115	-0.088	-0.063	-0.041
	0.25	-0.013	0.009	0.078	0.112	0.129	0.147	-0.153	-0.136	-0.120	-0.106	-0.095
	0.30	0.018	0.037	0.091	0.114	0.126	0.137	-0.163	-0.153	-0.144	-0.137	-0.132
	0.40	0.066	0.077	0.104	0.113	0.117	0.121	-0.179	-0.176	-0.174	-0.172	-0.172
	0.50	0.092	0.097	0.108	0.111	0.112	0.113	-0.187	-0.187	-0.187	-0.187	-0.188
	0.60	0.104	0.105	0.108	0.109	0.109	0.109	-0.191	-0.191	-0.191	-0.192	-0.192
	0.70	0.108	0.108	0.108	0.108	0.108	0.108	-0.192	-0.192	-0.192	-0.193	-0.193
	0.80	0.108	0.108	0.108	0.108	0.108	0.108	-0.192	-0.192	-0.192	-0.192	-0.192
$\bar{\sigma}_\theta / \eta_1$	0	-0.3	-0.3	-0.3	-0.3	-0.3	-0.3	-0.6	-0.6	-0.6	-0.6	-0.4
	0.01	-0.274	-0.273	-0.266	-0.262	-0.259	-0.257	-0.557	-0.540	-0.519	-0.474	-0.325
	0.02	-0.249	-0.246	-0.233	-0.225	-0.219	-0.211	-0.511	-0.484	-0.448	-0.388	-0.285
	0.05	-0.176	-0.169	-0.142	-0.123	-0.110	-0.094	-0.394	-0.356	-0.314	-0.267	-0.218
	0.10	-0.071	-0.062	-0.027	-0.008	0.004	0.017	-0.283	-0.252	-0.224	-0.197	-0.174
	0.15	0.003	0.011	0.037	0.050	0.056	0.063	-0.237	-0.213	-0.193	-0.175	-0.159
	0.20	0.047	0.053	0.068	0.074	0.077	0.079	-0.221	-0.201	-0.184	-0.170	-0.158
	0.25	0.069	0.072	0.078	0.080	0.081	0.082	-0.218	-0.200	-0.185	-0.173	-0.162
	0.30	0.077	0.078	0.079	0.079	0.079	0.079	-0.221	-0.204	-0.190	-0.178	-0.168
	0.40	0.074	0.073	0.071	0.070	0.069	0.069	-0.231	-0.215	-0.201	-0.189	-0.180
	0.50	0.065	0.064	0.063	0.062	0.061	0.061	-0.239	-0.222	-0.208	-0.197	-0.187
	0.60	0.059	0.059	0.058	0.057	0.057	0.057	-0.243	-0.226	-0.212	-0.200	-0.190
	0.70	0.056	0.056	0.055	0.055	0.055	0.055	-0.245	-0.228	-0.214	-0.202	-0.192
	0.80	0.054	0.054	0.054	0.054	0.054	0.054	-0.246	-0.229	-0.214	-0.202	-0.192
$\bar{\sigma}_r / \eta_1$	0	-0.3	-0.3	-0.3	-0.3	-0.3	-0.45	-0.45	-0.6	-0.6	-0.6	0
	0.01	-0.274	-0.271	-0.258	-0.253	-0.261	-0.332	-0.332	-0.396	-0.352	-0.201	0
	0.02	-0.249	-0.243	-0.216	-0.204	-0.207	-0.227	-0.227	-0.237	-0.175	-0.053	0
	0.05	-0.176	-0.162	-0.106	-0.074	-0.054	-0.030	-0.030	-0.016	0.003	0.011	0
	0.10	-0.071	-0.053	0.007	0.032	0.043	0.052	0.052	0.040	0.026	0.012	0
	0.15	0.003	0.017	0.050	0.058	0.060	0.059	0.059	0.041	0.024	0.010	0
	0.20	0.047	0.054	0.063	0.062	0.061	0.059	0.059	0.039	0.023	0.010	0
	0.25	0.069	0.070	0.066	0.062	0.060	0.057	0.057	0.038	0.022	0.010	0
	0.30	0.077	0.074	0.065	0.060	0.058	0.056	0.056	0.038	0.022	0.010	0
	0.40	0.074	0.070	0.061	0.058	0.056	0.055	0.055	0.037	0.022	0.010	0
	0.50	0.065	0.063	0.057	0.056	0.056	0.055	0.055	0.037	0.022	0.010	0
	0.60	0.059	0.058	0.055	0.055	0.055	0.054	0.054	0.037	0.022	0.010	0
	0.70	0.056	0.055	0.054	0.054	0.054	0.054	0.054	0.037	0.023	0.010	0
	0.80	0.054	0.054	0.054	0.054	0.054	0.054	0.054	0.037	0.023	0.010	0
$\bar{\tau}_{rz} / \eta_1$	0	0	0.042	0.073	0.061	0.027	$-\infty$	$-\infty$	0.108	0.217	0.303	0
	0.01	0	0.041	0.072	0.061	0.037	0.009	0.009	0.106	0.184	0.193	0
	0.02	0	0.040	0.068	0.060	0.048	0.050	0.050	0.089	0.115	0.066	0
	0.05	0	0.031	0.043	0.033	0.025	0.015	0.015	0.000	-0.021	-0.035	0
	0.10	0	0.007	-0.013	-0.031	-0.042	-0.051	-0.051	-0.056	-0.053	-0.036	0
	0.15	0	-0.015	-0.047	-0.058	-0.061	-0.061	-0.061	-0.056	-0.045	-0.026	0
	0.20	0	-0.027	-0.055	-0.058	-0.057	-0.053	-0.053	-0.045	-0.034	-0.019	0
	0.25	0	-0.030	-0.050	-0.049	-0.046	-0.041	-0.041	-0.034	-0.025	-0.013	0
	0.30	0	-0.027	-0.040	-0.037	-0.034	-0.030	-0.030	-0.024	-0.017	-0.009	0
	0.40	0	-0.016	-0.020	-0.018	-0.016	-0.014	-0.014	-0.011	-0.008	-0.004	0
	0.50	0	-0.007	-0.009	-0.007	-0.006	-0.005	-0.005	-0.004	-0.003	-0.001	0
	0.60	0	-0.003	-0.003	-0.002	-0.002	-0.002	-0.002	-0.001	0.001	0.000	0
	0.70	0	-0.001	-0.001	-0.001	0.000	0.000	0.000	0.000	0.000	0.000	0
	0.80	0	0.000	0.000	0.000	0.000	0.000	0.000	0.000	0.000	0.000	0

$\gamma = 0{,}5 ; \mu = 0{,}85 ; \delta = 1$

	λ	Core							Shell			
	p	0	0.3	0.6	0.7	0.75	0.80	0.85	0.85	0.90	0.95	1.00
$\dfrac{\bar\sigma_z}{\eta_1}$	0	0	0	0	0	0	0	0.15	-0.15	0	0	0.6
	0.01	-0.009	-0.009	-0.008	-0.003	0.005	0.027	0.119	-0.176	-0.084	-0.034	0.531
	0.02	-0.018	-0.019	-0.015	-0.005	0.008	0.038	0.103	-0.196	-0.117	0.017	0.457
	0.05	-0.043	-0.043	-0.029	-0.005	0.020	0.059	0.123	-0.178	-0.076	0.080	0.277
	0.10	-0.070	-0.064	-0.018	0.028	0.064	0.111	0.171	-0.129	-0.057	0.023	0.101
	0.15	-0.072	-0.058	0.011	0.061	0.093	0.130	0.171	-0.129	-0.086	-0.044	-0.005
	0.20	-0.055	-0.036	0.037	0.079	0.103	0.128	0.154	-0.146	-0.121	-0.097	-0.077
	0.25	-0.028	-0.009	0.056	0.087	0.103	0.119	0.135	-0.165	-0.150	-0.137	-0.127
	0.30	0.000	0.017	0.067	0.089	0.100	0.110	0.119	-0.181	-0.172	-0.165	-0.161
	0.40	0.044	0.054	0.079	0.088	0.092	0.095	0.098	-0.202	-0.200	-0.198	-0.198
	0.50	0.069	0.073	0.083	0.086	0.087	0.088	0.088	-0.212	-0.212	-0.212	-0.212
	0.60	0.079	0.081	0.083	0.084	0.084	0.085	0.084	-0.216	-0.216	-0.216	-0.217
	0.70	0.083	0.083	0.083	0.084	0.084	0.083	0.083	-0.217	-0.217	-0.217	-0.218
	0.80	0.083	0.083	0.083	0.083	0.083	0.083	0.083	-0.217	-0.217	-0.217	-0.217
$\dfrac{\bar\sigma_\theta}{\eta_1}$	0	-0.3	-0.3	-0.3	-0.3	-0.3	-0.3	-0.3	-0.6	-0.6	-0.6	-0.4
	0.01	-0.276	-0.274	-0.268	-0.263	-0.259	-0.254	-0.249	-0.549	-0.525	-0.480	-0.331
	0.02	-0.251	-0.248	-0.236	-0.226	-0.219	-0.210	-0.197	-0.497	-0.459	-0.399	-0.297
	0.05	-0.181	-0.174	-0.147	-0.128	-0.116	-0.099	-0.078	-0.378	-0.335	-0.289	-0.242
	0.10	-0.081	-0.071	-0.038	-0.019	-0.008	0.005	0.018	-0.282	-0.253	-0.226	-0.202
	0.15	-0.010	-0.002	0.024	0.036	0.042	0.048	0.054	-0.246	-0.223	-0.204	-0.188
	0.20	0.033	0.038	0.053	0.058	0.061	0.064	0.067	-0.233	-0.215	-0.199	-0.186
	0.25	0.054	0.057	0.063	0.065	0.066	0.067	0.068	-0.232	-0.216	-0.201	-0.190
	0.30	0.062	0.063	0.064	0.065	0.065	0.064	0.064	-0.236	-0.220	-0.206	-0.195
	0.40	0.060	0.059	0.057	0.056	0.056	0.055	0.055	-0.245	-0.230	-0.216	-0.205
	0.50	0.052	0.051	0.050	0.049	0.049	0.048	0.048	-0.252	-0.236	-0.223	-0.212
	0.60	0.046	0.046	0.045	0.045	0.044	0.044	0.044	-0.256	-0.240	-0.227	-0.215
	0.70	0.043	0.043	0.043	0.043	0.042	0.042	0.042	-0.258	-0.242	-0.228	-0.216
	0.80	0.042	0.042	0.042	0.042	0.042	0.042	0.042	-0.258	-0.242	-0.228	-0.217
$\dfrac{\bar\sigma_r}{\eta_1}$	0	-0.3	-0.3	-0.3	-0.3	-0.3	-0.3	-0.45	-0.45	-0.6	-0.6	0
	0.01	-0.276	-0.272	-0.258	-0.247	-0.241	-0.243	-0.297	-0.297	-0.331	-0.197	0
	0.02	-0.251	-0.245	-0.216	-0.196	-0.184	-0.175	-0.174	-0.175	-0.150	-0.048	0
	0.05	-0.181	-0.167	-0.109	-0.071	-0.048	-0.023	0.005	0.005	0.015	0.014	0
	0.10	-0.081	-0.062	-0.002	0.023	0.034	0.042	0.047	0.047	0.030	0.014	0
	0.15	-0.010	0.004	0.037	0.045	0.047	0.047	0.046	0.046	0.027	0.012	0
	0.20	0.033	0.040	0.050	0.049	0.048	0.046	0.044	0.044	0.026	0.011	0
	0.25	0.054	0.055	0.052	0.049	0.047	0.045	0.043	0.043	0.025	0.011	0
	0.30	0.062	0.060	0.052	0.047	0.045	0.044	0.042	0.042	0.025	0.011	0
	0.40	0.060	0.056	0.048	0.045	0.044	0.043	0.042	0.042	0.025	0.011	0
	0.50	0.052	0.050	0.046	0.043	0.043	0.042	0.042	0.042	0.025	0.012	0
	0.60	0.046	0.045	0.044	0.042	0.042	0.042	0.042	0.042	0.025	0.012	0
	0.70	0.043	0.043	0.042	0.042	0.042	0.042	0.042	0.042	0.025	0.012	0
	0.80	0.042	0.042	0.042	0.042	0.042	0.042	0.042	0.042	0.025	0.012	0
$\dfrac{\bar\tau_{rz}}{\eta_1}$	0	0	0.045	0.088	0.093	0.086	0.056	$-\infty$	$-\infty$	0.152	0.275	0
	0.01	0	0.045	0.086	0.090	0.084	0.063	0.041	0.041	0.131	0.167	0
	0.02	0	0.043	0.081	0.084	0.078	0.066	0.063	0.063	0.081	0.046	0
	0.05	0	0.034	0.052	0.044	0.033	0.017	-0.003	-0.003	-0.027	-0.038	0
	0.10	0	0.010	-0.007	-0.025	-0.036	-0.046	-0.052	-0.052	-0.049	-0.033	0
	0.15	0	-0.012	-0.041	-0.052	-0.055	-0.055	-0.051	-0.051	-0.041	-0.024	0
	0.20	0	-0.024	-0.050	-0.053	-0.052	-0.048	-0.041	-0.041	-0.031	-0.017	0
	0.25	0	-0.027	-0.045	-0.045	-0.042	-0.038	-0.031	-0.031	-0.023	-0.012	0
	0.30	0	-0.024	-0.037	-0.034	-0.031	-0.028	-0.022	-0.022	-0.016	-0.008	0
	0.40	0	-0.014	-0.019	-0.017	-0.015	-0.013	-0.010	-0.010	-0.007	-0.004	0
	0.50	0	-0.007	-0.008	-0.007	-0.006	-0.005	-0.004	-0.004	-0.003	-0.001	0
	0.60	0	-0.003	-0.003	-0.002	-0.002	-0.002	-0.001	-0.001	-0.001	0.000	0
	0.70	0	-0.001	-0.001	0.000	0.000	0.000	0.000	0.000	0.000	0.000	0
	0.80	0	0.000	0.000	0.000	0.000	0.000	0.000	0.000	0.000	0.000	0

	λ	Core								Shell		
	p	0	0.3	0.6	0.7	0.75	0.80	0.85	0.90	0.90	0.95	1.00
$\dfrac{\bar{\sigma}_s}{\eta_1}$	0	0	0	0	0	0	0	0	0.15	-0.15	0	0.6
	0.01	-0.010	-0.010	-0.010	-0.008	-0.006	0.001	0.022	0.118	-0.186	-0.068	0.493
	0.02	-0.019	-0.020	-0.020	-0.016	-0.010	0.002	0.034	0.111	-0.188	-0.033	0.392
	0.05	-0.045	-0.046	-0.038	-0.022	-0.005	0.026	0.080	0.173	-0.126	0.019	0.197
	0.10	-0.074	-0.070	-0.034	0.006	0.037	0.079	0.133	0.198	-0.102	-0.030	0.040
	0.15	-0.080	-0.068	-0.009	0.036	0.065	0.098	0.135	0.174	-0.126	-0.088	-0.052
	0.20	-0.067	-0.050	0.015	0.052	0.074	0.097	0.121	0.144	-0.156	-0.134	-0.115
	0.25	-0.044	-0.027	0.031	0.060	0.075	0.090	0.104	0.118	-0.182	-0.170	-0.160
	0.30	-0.019	-0.004	0.042	0.062	0.072	0.081	0.090	0.098	-0.202	-0.195	-0.191
	0.40	0.021	0.030	0.053	0.061	0.065	0.068	0.071	0.073	-0.227	-0.225	-0.225
	0.50	0.043	0.047	0.056	0.059	0.060	0.061	0.062	0.062	-0.238	-0.238	-0.239
	0.60	0.053	0.054	0.057	0.058	0.058	0.058	0.058	0.058	-0.242	-0.242	-0.243
	0.70	0.056	0.057	0.057	0.057	0.057	0.057	0.057	0.057	-0.243	-0.243	-0.243
	0.80	0.057	0.057	0.057	0.057	0.057	0.057	0.057	0.057	-0.243	-0.243	-0.243
$\dfrac{\bar{\sigma}_\theta}{\eta_1}$	0	-0.3	-0.3	-0.3	-0.3	-0.3	-0.3	-0.3	-0.3	-0.6	-0.6	-0.4
	0.01	-0.277	-0.276	-0.269	-0.264	-0.261	-0.256	-0.248	-0.235	-0.535	-0.488	-0.342
	0.02	-0.254	-0.251	-0.239	-0.230	-0.223	-0.213	-0.199	-0.176	-0.476	-0.415	-0.318
	0.05	-0.187	-0.181	-0.155	-0.137	-0.125	-0.109	-0.089	-0.063	-0.363	-0.317	-0.273
	0.10	-0.092	-0.083	-0.052	-0.034	-0.023	-0.011	0.001	0.013	-0.287	-0.259	-0.235
	0.15	-0.025	-0.017	0.007	0.018	0.024	0.030	0.036	0.042	-0.258	-0.237	-0.220
	0.20	0.016	0.022	0.035	0.041	0.044	0.046	0.049	0.051	-0.249	-0.231	-0.217
	0.25	0.038	0.040	0.046	0.049	0.049	0.050	0.051	0.052	-0.248	-0.232	-0.219
	0.30	0.046	0.047	0.048	0.049	0.049	0.049	0.049	0.048	-0.252	-0.236	-0.224
	0.40	0.044	0.044	0.043	0.042	0.041	0.041	0.040	0.040	-0.260	-0.245	-0.233
	0.50	0.038	0.037	0.036	0.035	0.035	0.035	0.034	0.034	-0.266	-0.251	-0.238
	0.60	0.033	0.032	0.032	0.031	0.031	0.031	0.031	0.031	-0.269	-0.254	-0.241
	0.70	0.031	0.030	0.030	0.029	0.029	0.029	0.029	0.029	-0.271	-0.256	-0.243
	0.80	0.029	0.029	0.029	0.029	0.029	0.029	0.029	0.029	-0.271	-0.256	-0.243
$\dfrac{\bar{\sigma}_r}{\eta_1}$	0	-0.3	-0.3	-0.3	-0.3	-0.3	-0.3	-0.3	-0.45	-0.45	-0.6	0
	0.01	-0.277	-0.274	-0.259	-0.247	-0.237	-0.225	-0.214	-0.242	-0.242	-0.182	0
	0.02	-0.254	-0.248	-0.219	-0.196	-0.179	-0.156	-0.130	-0.096	-0.096	-0.032	0
	0.05	-0.187	-0.174	-0.116	-0.076	-0.051	-0.023	0.005	0.028	0.028	0.018	0
	0.10	-0.092	-0.074	-0.016	0.010	0.021	0.029	0.033	0.034	0.034	0.015	0
	0.15	-0.025	-0.011	0.022	0.030	0.032	0.033	0.032	0.030	0.030	0.013	0
	0.20	0.016	0.023	0.034	0.034	0.033	0.032	0.030	0.029	0.029	0.013	0
	0.25	0.038	0.039	0.037	0.034	0.033	0.031	0.029	0.028	0.028	0.012	0
	0.30	0.046	0.044	0.037	0.034	0.032	0.030	0.029	0.028	0.028	0.013	0
	0.40	0.044	0.042	0.034	0.031	0.030	0.029	0.029	0.028	0.028	0.013	0
	0.50	0.038	0.036	0.031	0.030	0.029	0.029	0.029	0.029	0.029	0.013	0
	0.60	0.033	0.032	0.030	0.029	0.029	0.029	0.029	0.029	0.029	0.013	0
	0.70	0.031	0.030	0.029	0.029	0.029	0.029	0.029	0.029	0.029	0.013	0
	0.80	0.029	0.029	0.029	0.029	0.029	0.029	0.029	0.029	0.029	0.013	0
$\dfrac{\bar{\tau}_{rz}}{\eta_1}$	0	0	0.047	0.097	0.111	0.116	0.114	0.092	$-\infty$	$-\infty$	0.218	0
	0.01	0	0.047	0.096	0.109	0.112	0.109	0.091	0.069	0.069	0.123	0
	0.02	0	0.045	0.091	0.101	0.101	0.095	0.079	0.056	0.056	0.021	0
	0.05	0	0.037	0.061	0.055	0.045	0.028	0.004	-0.024	-0.024	-0.038	0
	0.10	0	0.013	0.001	-0.016	-0.026	-0.036	-0.043	-0.042	-0.042	-0.029	0
	0.15	0	-0.008	-0.033	-0.043	-0.047	-0.047	-0.044	-0.036	-0.036	-0.021	0
	0.20	0	-0.020	-0.043	-0.046	-0.045	-0.042	-0.036	-0.028	-0.028	-0.015	0
	0.25	0	-0.024	-0.040	-0.040	-0.037	-0.034	-0.028	-0.020	-0.020	-0.011	0
	0.30	0	-0.022	-0.033	-0.031	-0.028	-0.025	-0.020	-0.014	-0.014	-0.008	0
	0.40	0	-0.013	-0.017	-0.016	-0.014	-0.012	-0.009	-0.007	-0.007	-0.003	0
	0.50	0	-0.006	-0.008	-0.006	-0.006	-0.005	-0.004	-0.002	-0.002	-0.001	0
	0.60	0	-0.002	-0.003	-0.002	-0.002	-0.002	-0.001	-0.001	-0.001	0.000	0
	0.70	0	-0.001	-0.001	-0.001	0.000	0.000	0.000	0.000	0.000	0.000	0
	0.80	0	0.000	0.000	0.000	0.000	0.000	0.000	0.000	0.000	0.000	0

$\gamma = -0,5;\ \mu = 0,95;\ \delta = 1$

	λ / p	Core									Shell	
		0	0.3	0.6	0.7	0.75	0.80	0.85	0.90	0.95	0.95	1.00
$\bar{\sigma}_z / \eta_1$	0	0	0	0	0	0	0	0	0	0.15	-0.15	0.6
	0.01	-0.010	-0.010	-0.011	-0.011	-0.010	-0.010	-0.005	0.014	0.124	-0.178	0.393
	0.02	-0.020	-0.020	-0.022	-0.021	-0.019	-0.015	-0.003	0.036	0.174	-0.128	0.259
	0.05	-0.046	-0.047	-0.045	-0.035	-0.024	-0.001	0.040	0.116	0.239	-0.062	0.089
	0.10	-0.077	-0.075	-0.047	-0.015	0.011	0.045	0.091	0.146	0.209	-0.091	-0.028
	0.15	-0.087	-0.078	-0.028	0.010	0.034	0.063	0.095	0.130	0.165	-0.135	-0.102
	0.20	-0.079	-0.065	-0.009	0.024	0.043	0.064	0.085	0.106	0.126	-0.174	-0.155
	0.25	-0.060	-0.045	0.006	0.031	0.044	0.058	0.071	0.084	0.096	-0.204	-0.194
	0.30	-0.039	-0.025	0.015	0.033	0.042	0.051	0.059	0.067	0.073	-0.227	-0.222
	0.40	-0.004	0.004	0.025	0.033	0.036	0.040	0.042	0.044	0.046	-0.254	-0.253
	0.50	0.016	0.020	0.029	0.032	0.033	0.033	0.034	0.034	0.034	-0.266	-0.266
	0.60	0.025	0.027	0.029	0.031	0.030	0.030	0.030	0.030	0.030	-0.270	-0.270
	0.70	0.029	0.029	0.029	0.030	0.029	0.029	0.029	0.029	0.029	-0.271	-0.271
	0.80	0.029	0.029	0.029	0.029	0.029	0.029	0.029	0.029	0.029	-0.271	-0.271
$\bar{\sigma}_\theta / \eta_1$	0	-0.3	-0.3	-0.3	-0.3	-0.3	-0.3	-0.3	-0.3	-0.3	-0.6	-0.4
	0.01	-0.278	-0.277	-0.272	-0.267	-0.264	-0.259	-0.252	-0.238	-0.205	-0.505	-0.366
	0.02	-0.257	-0.254	-0.244	-0.235	-0.228	-0.220	-0.207	-0.184	-0.142	-0.442	-0.354
	0.05	-0.195	-0.190	-0.166	-0.149	-0.138	-0.124	-0.105	-0.082	-0.055	-0.355	-0.313
	0.10	-0.106	-0.098	-0.069	-0.052	-0.043	-0.032	-0.021	-0.009	0.002	-0.298	-0.272
	0.15	-0.042	-0.035	-0.012	-0.002	0.004	0.010	0.015	0.021	0.026	-0.274	-0.255
	0.20	-0.002	0.002	0.016	0.021	0.024	0.027	0.029	0.032	0.034	-0.266	-0.250
	0.25	0.019	0.021	0.028	0.030	0.031	0.032	0.033	0.034	0.034	-0.266	-0.251
	0.30	0.028	0.029	0.031	0.031	0.032	0.032	0.032	0.032	0.032	-0.268	-0.254
	0.40	0.028	0.028	0.027	0.026	0.026	0.026	0.025	0.025	0.024	-0.276	-0.261
	0.50	0.023	0.023	0.021	0.021	0.020	0.020	0.020	0.020	0.019	-0.281	-0.266
	0.60	0.019	0.018	0.018	0.017	0.017	0.017	0.017	0.017	0.016	-0.284	-0.269
	0.70	0.016	0.016	0.016	0.016	0.015	0.015	0.015	0.015	0.015	-0.285	-0.270
	0.80	0.015	0.015	0.015	0.015	0.015	0.015	0.015	0.015	0.015	-0.285	-0.271
$\bar{\sigma}_r / \eta_1$	0	-0.3	-0.3	-0.3	-0.3	-0.3	-0.3	-0.3	-0.3	-0.45	-0.45	0
	0.01	-0.278	-0.276	-0.262	-0.250	-0.241	-0.226	-0.203	-0.164	-0.108	-0.108	0
	0.02	-0.257	-0.251	-0.225	-0.202	-0.185	-0.159	-0.122	-0.067	-0.002	-0.002	0
	0.05	-0.195	-0.182	-0.128	-0.090	-0.065	-0.037	-0.008	0.015	0.020	0.020	0
	0.10	-0.106	-0.089	-0.034	-0.009	0.002	0.011	0.017	0.018	0.016	0.016	0
	0.15	-0.042	-0.029	0.003	0.012	0.015	0.017	0.017	0.016	0.015	0.015	0
	0.20	-0.002	0.005	0.017	0.018	0.018	0.017	0.016	0.015	0.014	0.014	0
	0.25	0.019	0.021	0.021	0.019	0.017	0.016	0.015	0.014	0.014	0.014	0
	0.30	0.028	0.027	0.021	0.019	0.017	0.016	0.015	0.014	0.014	0.014	0
	0.40	0.028	0.026	0.020	0.017	0.016	0.015	0.015	0.014	0.014	0.014	0
	0.50	0.023	0.021	0.017	0.016	0.015	0.015	0.015	0.015	0.014	0.014	0
	0.60	0.019	0.018	0.016	0.015	0.015	0.015	0.015	0.015	0.015	0.015	0
	0.70	0.016	0.016	0.015	0.015	0.015	0.015	0.015	0.015	0.015	0.015	0
	0.80	0.015	0.015	0.015	0.015	0.015	0.015	0.015	0.015	0.015	0.015	0
$\bar{\tau}_{rz} / \eta_1$	0	0	0.048	0.102	0.122	0.131	0.141	0.146	0.136	$-\infty$	$-\infty$	0
	0.01	0	0.048	0.101	0.119	0.127	0.134	0.135	0.119	0.065	0.065	0
	0.02	0	0.047	0.096	0.111	0.116	0.117	0.108	0.078	0.010	0.010	0
	0.05	0	0.038	0.068	0.066	0.059	0.044	0.021	-0.008	-0.026	-0.026	0
	0.10	0	0.017	0.011	-0.003	-0.012	-0.022	-0.029	-0.031	-0.022	-0.022	0
	0.15	0	-0.004	-0.023	-0.033	-0.036	-0.037	-0.035	-0.029	-0.017	-0.017	0
	0.20	0	-0.016	-0.035	-0.038	-0.037	-0.035	-0.031	-0.023	-0.013	-0.013	0
	0.25	0	-0.020	-0.034	-0.034	-0.032	-0.029	-0.024	-0.018	-0.010	-0.010	0
	0.30	0	-0.019	-0.029	-0.027	-0.025	-0.022	-0.018	-0.013	-0.007	-0.007	0
	0.40	0	-0.012	-0.016	-0.014	-0.013	-0.011	-0.009	-0.006	-0.003	-0.003	0
	0.50	0	-0.006	-0.007	-0.006	-0.005	-0.004	-0.003	-0.002	-0.001	-0.001	0
	0.60	0	-0.002	-0.003	-0.002	-0.002	-0.001	-0.001	-0.001	0.000	0.000	0
	0.70	0	-0.001	-0.001	-0.001	0.000	0.000	0.000	0.000	0.000	0.000	0
	0.80	0	0.000	0.000	0.000	0.000	0.000	0.000	0.000	0.000	0.000	0

	λ					Continuous Homogeneous Cylinder					
	p	0	0.3	0.6	0.7	0.75	0.80	0.85	0.90	0.95	1.00
	0	0	0	0	0	0	0	0	0	0	0.6
	0.01	−0.019	−0.020	−0.023	−0.024	−0.025	−0.025	−0.025	−0.020	0.018	0.593
	0.02	−0.039	−0.040	−0.045	−0.047	−0.046	−0.045	−0.037	−0.009	0.113	0.576
	0.05	−0.092	−0.095	−0.096	−0.086	−0.073	−0.046	0.005	0.102	0.272	0.511
	0.10	−0.158	−0.155	−0.118	−0.073	−0.035	0.017	0.087	0.176	0.281	0.395
	0.15	−0.185	−0.172	−0.096	−0.037	0.003	0.050	0.105	0.165	0.228	0.291
$\bar\sigma_z$	0.20	−0.180	−0.158	−0.067	−0.012	0.020	0.056	0.093	0.132	0.170	0.206
η_1	0.25	−0.154	−0.129	−0.043	0.000	0.024	0.049	0.073	0.097	0.120	0.140
	0.30	−0.121	−0.097	−0.026	0.005	0.022	0.038	0.054	0.068	0.081	0.091
	0.40	−0.061	−0.047	−0.008	0.006	0.013	0.019	0.024	0.029	0.032	0.034
	0.50	−0.025	−0.018	−0.002	0.004	0.006	0.008	0.009	0.010	0.010	0.010
	0.60	−0.008	−0.006	0.000	0.002	0.002	0.003	0.003	0.002	0.002	0.001
	0.70	−0.002	−0.001	0.000	0.001	0.001	0.001	0.000	0.000	0.000	0.000
	0.80	0.000	0.000	0.000	0.000	0.000	0.000	0.000	0.000	0.000	0.000
	0	−0.6	−0.6	−0.6	−0.6	−0.6	−0.6	−0.6	−0.6	−0.6	−0.4
	0.01	−0.561	−0.559	−0.549	−0.542	−0.537	−0.530	−0.519	−0.499	−0.454	−0.305
	0.02	−0.522	−0.518	−0.499	−0.486	−0.476	−0.462	−0.442	−0.409	−0.350	−0.245
	0.05	−0.409	−0.399	−0.360	−0.332	−0.314	−0.290	−0.261	−0.223	−0.178	−0.129
	0.10	−0.246	−0.231	−0.180	−0.151	−0.134	−0.115	−0.095	−0.073	−0.051	−0.030
	0.15	−0.125	−0.112	−0.071	−0.051	−0.041	−0.030	−0.018	−0.007	0.003	0.013
$\bar\sigma_\theta$	0.20	−0.048	−0.039	−0.013	−0.002	0.004	0.009	0.015	0.020	0.025	0.029
η_1	0.25	−0.004	0.001	0.014	0.019	0.022	0.024	0.026	0.028	0.030	0.032
	0.30	0.016	0.018	0.024	0.025	0.026	0.027	0.027	0.028	0.028	0.028
	0.40	0.022	0.022	0.021	0.020	0.020	0.019	0.019	0.018	0.017	0.017
	0.50	0.015	0.014	0.012	0.011	0.010	0.010	0.009	0.009	0.008	0.008
	0.60	0.007	0.007	0.005	0.005	0.005	0.004	0.004	0.004	0.003	0.003
	0.70	0.003	0.003	0.002	0.002	0.002	0.001	0.001	0.001	0.001	0.001
	0.80	0.000	0.000	0.000	0.000	0.000	0.000	0.000	0.000	0.000	0.000
	0	−0.6	−0.6	−0.6	−0.6	−0.6	−0.6	−0.6	−0.6	−0.6	0
	0.01	−0.561	−0.556	−0.533	−0.513	−0.498	−0.474	−0.436	−0.366	−0.205	0
	0.02	−0.522	−0.512	−0.467	−0.430	−0.402	−0.360	−0.298	−0.200	−0.060	0
	0.05	−0.409	−0.387	−0.295	−0.230	−0.186	−0.136	−0.080	−0.029	0.000	0
	0.10	−0.246	−0.216	−0.115	−0.068	−0.045	−0.024	−0.008	0.000	0.002	0
	0.15	−0.125	−0.101	−0.037	−0.016	−0.008	−0.003	0.000	0.001	0.000	0
$\bar\sigma_r$	0.20	−0.048	−0.034	−0.005	0.000	0.002	0.002	0.001	0.000	−0.001	0
η_1	0.25	−0.004	0.001	0.006	0.005	0.003	0.002	0.001	0.000	−0.001	0
	0.30	0.016	0.016	0.010	0.006	0.004	0.002	0.001	0.000	−0.001	0
	0.40	0.022	0.018	0.008	0.004	0.003	0.001	0.000	0.000	0.000	0
	0.50	0.015	0.012	0.005	0.002	0.001	0.001	0.000	0.000	0.000	0
	0.60	0.007	0.006	0.002	0.001	0.001	0.000	0.000	0.000	0.000	0
	0.70	0.003	0.002	0.001	0.000	0.000	0.000	0.000	0.000	0.000	0
	0.80	0.000	0.000	0.000	0.000	0.000	0.000	0.000	0.000	0.000	0
	0	0	0.095	0.204	0.246	0.268	0.293	0.317	0.342	0.367	0
	0.01	0	0.094	0.201	0.241	0.261	0.282	0.299	0.305	0.254	0
	0.02	0	0.092	0.193	0.228	0.243	0.254	0.254	0.226	0.122	0
	0.05	0	0.078	0.145	0.152	0.146	0.128	0.096	0.048	0.001	0
	0.10	0	0.041	0.044	0.025	0.011	−0.006	−0.021	−0.030	−0.025	0
	0.15	0	0.003	−0.022	−0.038	−0.045	−0.048	−0.048	−0.041	−0.025	0
$\bar\tau_{rz}$	0.20	0	−0.020	−0.050	−0.057	−0.057	−0.054	−0.048	−0.037	−0.021	0
η_1	0.25	0	−0.030	−0.055	−0.055	−0.053	−0.048	−0.041	−0.030	−0.017	0
	0.30	0	−0.031	−0.048	−0.046	−0.043	−0.038	−0.031	−0.023	−0.012	0
	0.40	0	−0.021	−0.028	−0.025	−0.023	−0.020	−0.016	−0.011	−0.006	0
	0.50	0	−0.010	−0.013	−0.011	−0.010	−0.008	−0.007	−0.004	−0.002	0
	0.60	0	−0.004	−0.005	−0.004	−0.003	−0.003	−0.002	−0.001	−0.001	0
	0.70	0	−0.001	−0.001	−0.001	−0.001	−0.001	−0.001	0.000	0.000	0
	0.80	0	0.000	0.000	0.000	0.000	0.000	0.000	0.000	0.000	0

$\gamma = 1.5;\ \mu = 0.7;\ \delta = 1$

	λ / p	Core				Shell						
		0	0.3	0.6	0.7	0.7	0.75	0.80	0.85	0.90	0.95	1.00
$\dfrac{\bar{\sigma}_x}{\eta_1}$	0	0	0	0	-0.15	0.15	0	0	0	0	0	0.6
	0.01	-0.032	-0.035	-0.059	-0.181	0.118	0.024	0.002	-0.004	-0.001	0.037	0.617
	0.02	-0.063	-0.069	-0.111	-0.209	0.093	0.032	0.006	0.003	0.028	0.152	0.622
	0.05	-0.152	-0.163	-0.216	-0.253	0.046	0.037	0.047	0.089	0.185	0.362	0.615
	0.10	-0.267	-0.274	-0.272	-0.241	0.058	0.091	0.141	0.212	0.307	0.424	0.556
	0.15	-0.328	-0.321	-0.260	-0.202	0.098	0.139	0.190	0.250	0.319	0.394	0.472
	0.20	-0.342	-0.322	-0.232	-0.173	0.127	0.163	0.204	0.248	0.294	0.342	0.388
	0.25	-0.325	-0.299	-0.206	-0.157	0.143	0.171	0.201	0.231	0.261	0.290	0.317
	0.30	-0.293	-0.267	-0.187	-0.149	0.151	0.171	0.191	0.211	0.230	0.247	0.262
	0.40	-0.229	-0.211	-0.165	-0.146	0.153	0.162	0.170	0.178	0.184	0.189	0.193
	0.50	-0.186	-0.177	-0.156	-0.149	0.151	0.154	0.157	0.159	0.161	0.162	0.162
	0.60	-0.165	-0.161	-0.153	-0.151	0.149	0.150	0.151	0.151	0.151	0.151	0.150
	0.70	-0.156	-0.155	-0.153	-0.152	0.148	0.148	0.148	0.148	0.147	0.147	0.147
	0.80	-0.153	-0.153	-0.153	-0.153	0.147	0.147	0.147	0.147	0.147	0.147	0.147
$\dfrac{\bar{\sigma}_\theta}{\eta_1}$	0	-0.9	-0.9	-0.9	-0.9	-0.6	-0.6	-0.6	-0.6	-0.6	-0.6	-0.4
	0.01	-0.849	-0.846	-0.832	-0.818	-0.518	-0.520	-0.515	-0.505	-0.486	-0.442	-0.293
	0.02	-0.798	-0.792	-0.766	-0.742	-0.442	-0.442	-0.433	-0.415	-0.384	-0.325	-0.221
	0.05	-0.649	-0.636	-0.582	-0.544	-0.244	-0.238	-0.223	-0.198	-0.164	-0.120	-0.071
	0.10	-0.434	-0.415	-0.347	-0.309	-0.009	-0.007	0.002	0.015	0.031	0.050	0.069
	0.15	-0.272	-0.255	-0.200	-0.174	0.126	0.121	0.120	0.122	0.126	0.131	0.137
	0.20	-0.165	-0.153	-0.117	-0.102	0.198	0.187	0.179	0.174	0.170	0.168	0.167
	0.25	-0.102	-0.095	-0.075	-0.067	0.233	0.217	0.205	0.196	0.188	0.182	0.177
	0.30	-0.070	-0.066	-0.057	-0.054	0.246	0.228	0.214	0.202	0.192	0.184	0.177
	0.40	-0.054	-0.054	-0.054	-0.054	0.246	0.226	0.210	0.196	0.185	0.175	0.167
	0.50	-0.060	-0.060	-0.062	-0.063	0.237	0.217	0.200	0.187	0.175	0.165	0.157
	0.60	-0.068	-0.068	-0.070	-0.070	0.230	0.210	0.194	0.180	0.169	0.159	0.151
	0.70	-0.073	-0.073	-0.074	-0.074	0.226	0.206	0.190	0.177	0.166	0.156	0.148
	0.80	-0.076	-0.076	-0.076	-0.076	0.224	0.204	0.188	0.176	0.165	0.155	0.147
$\dfrac{\bar{\sigma}_r}{\eta_1}$	0	-0.9	-0.9	-0.9	-0.75	-0.75	-0.6	-0.6	-0.6	-0.6	-0.6	0
	0.01	-0.849	-0.842	-0.798	-0.663	-0.663	-0.544	-0.495	-0.446	-0.370	-0.207	0
	0.02	-0.798	-0.784	-0.702	-0.575	-0.575	-0.474	-0.397	-0.316	-0.208	-0.063	0
	0.05	-0.649	-0.619	-0.472	-0.362	-0.362	-0.276	-0.193	-0.113	-0.045	-0.006	0
	0.10	-0.434	-0.392	-0.246	-0.176	-0.176	-0.123	-0.078	-0.042	-0.018	-0.005	0
	0.15	-0.272	-0.237	-0.142	-0.109	-0.109	-0.077	-0.051	-0.031	-0.017	-0.007	0
	0.20	-0.165	-0.144	-0.096	-0.085	-0.085	-0.061	-0.043	-0.029	-0.017	-0.008	0
	0.25	-0.102	-0.093	-0.077	-0.075	-0.075	-0.056	-0.040	-0.028	-0.018	-0.009	0
	0.30	-0.070	-0.068	-0.069	-0.072	-0.072	-0.054	-0.040	-0.028	-0.018	-0.009	0
	0.40	-0.054	-0.057	-0.068	-0.072	-0.072	-0.054	-0.040	-0.028	-0.018	-0.009	0
	0.50	-0.060	-0.063	-0.071	-0.074	-0.074	-0.055	-0.041	-0.028	-0.017	-0.008	0
	0.60	-0.068	-0.069	-0.074	-0.075	-0.075	-0.056	-0.041	-0.028	-0.017	-0.008	0
	0.70	-0.073	-0.074	-0.075	-0.076	-0.076	-0.057	-0.041	-0.028	-0.017	-0.008	0
	0.80	-0.076	-0.076	-0.076	-0.076	-0.076	-0.057	-0.041	-0.028	-0.017	-0.008	0
$\dfrac{\bar{\tau}_{rz}}{\eta_1}$	0	0	0.158	0.389	$+\infty$	$+\infty$	0.492	0.443	0.422	0.411	0.402	0
	0.01	0	0.157	0.381	0.522	0.522	0.470	0.428	0.402	0.373	0.289	0
	0.02	0	0.154	0.361	0.440	0.440	0.422	0.389	0.352	0.290	0.155	0
	0.05	0	0.134	0.267	0.277	0.277	0.259	0.223	0.169	0.097	0.026	0
	0.10	0	0.080	0.109	0.085	0.085	0.064	0.039	0.014	-0.006	-0.013	0
	0.15	0	0.027	0.009	-0.011	-0.011	-0.022	-0.030	-0.035	-0.033	-0.022	0
	0.20	0	-0.009	-0.039	-0.049	-0.049	-0.051	-0.050	-0.046	-0.037	-0.022	0
	0.25	0	-0.027	-0.054	-0.056	-0.056	-0.054	-0.050	-0.043	-0.033	-0.018	0
	0.30	0	-0.032	-0.052	-0.051	-0.051	-0.048	-0.043	-0.035	-0.026	-0.014	0
	0.40	0	-0.024	-0.033	-0.030	-0.030	-0.028	-0.024	-0.019	-0.014	-0.007	0
	0.50	0	-0.013	-0.016	-0.014	-0.014	-0.013	-0.011	-0.008	-0.006	-0.003	0
	0.60	0	-0.006	-0.007	-0.006	-0.006	-0.005	-0.004	-0.003	-0.002	-0.001	0
	0.70	0	-0.002	-0.002	-0.002	-0.002	-0.001	-0.001	-0.001	-0.001	0.000	0
	0.80	0	0.000	0.000	0.000	0.000	0.000	0.000	0.000	0.000	0.000	0

	λ	Core					Shell					
	p	0	0.3	0.6	0.7	0.75	0.75	0.80	0.85	0.90	0.95	1.00
$\dfrac{\bar\sigma_z}{\eta_1}$	0	0	0	0	0	−0.15	0.15	0	0	0	0	0.6
	0.01	−0.031	−0.033	−0.048	−0.083	−0.180	0.121	0.027	0.006	0.006	0.044	0.625
	0.02	−0.061	−0.066	−0.093	−0.141	−0.204	0.098	0.039	0.021	0.041	0.164	0.637
	0.05	−0.148	−0.156	−0.194	−0.220	−0.230	0.071	0.076	0.114	0.209	0.387	0.646
	0.10	−0.259	−0.263	−0.253	−0.219	−0.185	0.115	0.166	0.239	0.336	0.456	0.593
	0.15	−0.318	−0.309	−0.241	−0.181	−0.138	0.162	0.215	0.277	0.348	0.425	0.506
	0.20	−0.329	−0.308	−0.213	−0.151	−0.114	0.186	0.228	0.274	0.322	0.371	0.419
	0.25	−0.310	−0.283	−0.186	−0.135	−0.106	0.194	0.224	0.256	0.287	0.317	0.345
	0.30	−0.277	−0.250	−0.166	−0.127	−0.107	0.193	0.214	0.235	0.254	0.272	0.287
	0.40	−0.210	−0.191	−0.143	−0.125	−0.116	0.184	0.193	0.201	0.207	0.212	0.216
	0.50	−0.165	−0.156	−0.134	−0.127	−0.124	0.176	0.179	0.181	0.183	0.184	0.184
	0.60	−0.143	−0.140	−0.131	−0.129	−0.128	0.172	0.172	0.173	0.173	0.172	0.172
	0.70	−0.134	−0.133	−0.131	−0.130	−0.130	0.170	0.170	0.169	0.169	0.169	0.168
	0.80	−0.131	−0.131	−0.131	−0.131	−0.131	0.169	0.169	0.169	0.169	0.169	0.169
$\dfrac{\bar\sigma_\theta}{\eta_1}$	0	−0.9	−0.9	−0.9	−0.9	−0.9	−0.6	−0.6	−0.6	−0.6	−0.6	−0.4
	0.01	−0.848	−0.845	−0.833	−0.822	−0.811	−0.511	−0.511	−0.502	−0.483	−0.439	−0.289
	0.02	−0.796	−0.791	−0.766	−0.747	−0.730	−0.430	−0.425	−0.409	−0.378	−0.319	−0.214
	0.05	−0.646	−0.633	−0.581	−0.544	−0.519	−0.219	−0.207	−0.184	−0.150	−0.107	−0.057
	0.10	−0.428	−0.409	−0.342	−0.304	−0.281	0.019	0.025	0.037	0.053	0.070	0.089
	0.15	−0.264	−0.246	−0.191	−0.164	−0.150	0.150	0.147	0.148	0.150	0.155	0.160
	0.20	−0.155	−0.142	−0.106	−0.090	−0.082	0.218	0.208	0.201	0.196	0.193	0.190
	0.25	−0.091	−0.083	−0.063	−0.055	−0.052	0.248	0.234	0.223	0.214	0.206	0.200
	0.30	−0.058	−0.054	−0.045	−0.042	−0.040	0.260	0.243	0.229	0.218	0.208	0.200
	0.40	−0.042	−0.042	−0.042	−0.042	−0.043	0.257	0.239	0.223	0.210	0.199	0.189
	0.50	−0.048	−0.049	−0.051	−0.052	−0.053	0.247	0.229	0.213	0.200	0.189	0.179
	0.60	−0.056	−0.057	−0.059	−0.059	−0.060	0.240	0.222	0.206	0.193	0.182	0.173
	0.70	−0.062	−0.062	−0.063	−0.063	−0.063	0.237	0.218	0.203	0.190	0.179	0.170
	0.80	−0.065	−0.065	−0.065	−0.065	−0.065	0.235	0.216	0.201	0.189	0.178	0.169
$\dfrac{\bar\sigma_r}{\eta_1}$	0	−0.9	−0.9	−0.9	−0.9	−0.75	−0.75	−0.6	−0.6	−0.6	−0.6	0
	0.01	−0.848	−0.841	−0.805	−0.753	−0.644	−0.644	−0.519	−0.454	−0.373	−0.207	0
	0.02	−0.796	−0.783	−0.714	−0.631	−0.541	−0.541	−0.428	−0.329	−0.213	−0.064	0
	0.05	−0.646	−0.616	−0.481	−0.376	−0.307	−0.307	−0.215	−0.126	−0.052	−0.008	0
	0.10	−0.428	−0.386	−0.243	−0.173	−0.138	−0.138	−0.089	−0.049	−0.022	−0.007	0
	0.15	−0.264	−0.229	−0.134	−0.100	−0.087	−0.087	−0.058	−0.036	−0.020	−0.009	0
	0.20	−0.155	−0.133	−0.086	−0.074	−0.070	−0.070	−0.049	−0.033	−0.020	−0.010	0
	0.25	−0.091	−0.081	−0.066	−0.064	−0.064	−0.064	−0.046	−0.032	−0.020	−0.010	0
	0.30	−0.058	−0.056	−0.058	−0.061	−0.062	−0.062	−0.046	−0.032	−0.020	−0.010	0
	0.40	−0.042	−0.046	−0.056	−0.061	−0.062	−0.062	−0.046	−0.032	−0.020	−0.010	0
	0.50	−0.048	−0.052	−0.060	−0.063	−0.064	−0.064	−0.047	−0.032	−0.020	−0.009	0
	0.60	−0.056	−0.058	−0.063	−0.064	−0.065	−0.065	−0.047	−0.032	−0.020	−0.009	0
	0.70	−0.062	−0.063	−0.064	−0.065	−0.065	−0.065	−0.047	−0.032	−0.020	−0.009	0
	0.80	−0.065	−0.065	−0.065	−0.065	−0.065	−0.065	−0.047	−0.032	−0.020	−0.009	0
$\dfrac{\bar\tau_{rz}}{\eta_1}$	0	0	0.153	0.357	0.491	$+\infty$	$+\infty$	0.512	0.459	0.433	0.413	0
	0.01	0	0.152	0.352	0.470	0.536	0.536	0.486	0.437	0.394	0.300	0
	0.02	0	0.149	0.337	0.425	0.453	0.453	0.429	0.381	0.310	0.165	0
	0.05	0	0.130	0.258	0.277	0.267	0.267	0.236	0.181	0.107	0.032	0
	0.10	0	0.077	0.107	0.085	0.065	0.065	0.041	0.015	−0.005	−0.013	0
	0.15	0	0.025	0.006	−0.014	−0.024	−0.024	−0.032	−0.036	−0.034	−0.023	0
	0.20	0	−0.011	−0.041	−0.051	−0.053	−0.053	−0.053	−0.048	−0.038	−0.023	0
	0.25	0	−0.029	−0.056	−0.059	−0.057	−0.057	−0.052	−0.045	−0.034	−0.019	0
	0.30	0	−0.033	−0.054	−0.053	−0.050	−0.050	−0.044	−0.037	−0.027	−0.015	0
	0.40	0	−0.025	−0.035	−0.032	−0.029	−0.029	−0.025	−0.020	−0.014	−0.007	0
	0.50	0	−0.013	−0.017	−0.015	−0.013	−0.013	−0.011	−0.009	−0.006	−0.003	0
	0.60	0	−0.006	−0.007	−0.006	−0.005	−0.005	−0.004	−0.003	−0.002	−0.001	0
	0.70	0	−0.002	−0.002	−0.002	−0.001	−0.001	−0.001	−0.001	−0.001	0.000	0
	0.80	0	0.000	0.000	0.000	0.000	0.000	0.000	0.000	0.000	0.000	0

γ = 0,5; μ = 0,8; δ = 1

	λ → p ↓	Core						Shell				
		0	0.3	0.6	0.7	0.75	0.80	0.80	0.85	0.90	0.95	1.00
$\bar{\sigma}_z / \eta_1$	0	0	0	0	0	0	-0.15	0.1=	0	0	0	0.6
	0.01	-0.030	-0.032	-0.042	-0.057	-0.081	-0.178	0.124	0.031	0.017	0.053	0.637
	0.02	-0.060	-0.064	-0.082	-0.106	-0.136	-0.197	0.106	0.055	0.061	0.182	0.660
	0.05	-0.144	-0.151	-0.176	-0.191	-0.195	-0.189	0.112	0.147	0.240	0.420	0.687
	0.10	-0.252	-0.254	-0.235	-0.196	-0.160	-0.106	0.193	0.269	0.370	0.495	0.637
	0.15	-0.308	-0.297	-0.222	-0.158	-0.113	-0.059	0.242	0.306	0.380	0.460	0.544
	0.20	-0.316	-0.293	-0.192	-0.128	-0.089	-0.045	0.254	0.302	0.352	0.402	0.452
	0.25	-0.295	-0.266	-0.165	-0.111	-0.081	-0.050	0.250	0.283	0.315	0.346	0.374
	0.30	-0.259	-0.231	-0.144	-0.104	-0.082	-0.061	0.239	0.260	0.281	0.299	0.314
	0.40	-0.189	-0.170	-0.120	-0.101	-0.092	-0.083	0.217	0.225	0.232	0.237	0.240
	0.50	-0.143	-0.134	-0.111	-0.103	-0.100	-0.097	0.203	0.205	0.207	0.207	0.207
	0.60	-0.120	-0.116	-0.108	-0.106	-0.105	-0.104	0.196	0.196	0.196	0.196	0.195
	0.70	-0.111	-0.110	-0.108	-0.107	-0.107	-0.107	0.193	0.193	0.192	0.192	0.191
	0.80	-0.108	-0.108	-0.108	-0.108	-0.108	-0.108	0.192	0.192	0.192	0.192	0.192
$\bar{\sigma}_\theta / \eta_1$	0	-0.9	-0.9	-0.9	-0.9	-0.9	-0.9	-0.6	-0.6	-0.6	-0.6	-0.4
	0.01	-0.847	-0.844	-0.832	-0.823	-0.815	-0.802	-0.502	-0.497	-0.479	-0.435	-0.284
	0.02	-0.795	-0.789	-0.765	-0.747	-0.733	-0.713	-0.413	-0.400	-0.370	-0.311	-0.206
	0.05	-0.642	-0.629	-0.578	-0.542	-0.517	-0.487	-0.187	-0.166	-0.133	-0.090	-0.039
	0.10	-0.420	-0.401	-0.334	-0.295	-0.272	-0.247	0.053	0.063	0.077	0.095	0.113
	0.15	-0.253	-0.235	-0.180	-0.152	-0.138	-0.122	0.178	0.176	0.178	0.181	0.185
	0.20	-0.143	-0.130	-0.093	-0.077	-0.069	-0.061	0.239	0.230	0.224	0.219	0.216
	0.25	-0.078	-0.070	-0.050	-0.042	-0.038	-0.034	0.266	0.253	0.242	0.233	0.225
	0.30	-0.045	-0.041	-0.032	-0.029	-0.027	-0.026	0.274	0.258	0.245	0.234	0.225
	0.40	-0.029	-0.029	-0.029	-0.030	-0.030	-0.031	0.269	0.252	0.237	0.224	0.213
	0.50	-0.036	-0.037	-0.039	-0.040	-0.040	-0.041	0.259	0.241	0.226	0.214	0.203
	0.60	-0.044	-0.045	-0.047	-0.047	-0.048	-0.048	0.252	0.234	0.220	0.207	0.196
	0.70	-0.050	-0.050	-0.051	-0.051	-0.052	-0.052	0.248	0.231	0.216	0.204	0.193
	0.80	-0.054	-0.054	-0.054	-0.054	-0.054	-0.054	0.246	0.229	0.215	0.203	0.192
$\bar{\sigma}_r / \eta_1$	0	-0.9	-0.9	-0.9	-0.9	-0.9	-0.75	-0.75	-0.6	-0.6	-0.6	0
	0.01	-0.847	-0.841	-0.808	-0.774	-0.734	-0.616	-0.616	-0.477	-0.380	-0.209	0
	0.02	-0.795	-0.782	-0.718	-0.656	-0.596	-0.493	-0.493	-0.359	-0.224	-0.067	0
	0.05	-0.642	-0.613	-0.483	-0.386	-0.320	-0.242	-0.242	-0.145	-0.061	-0.011	0
	0.10	-0.420	-0.379	-0.238	-0.167	-0.132	-0.100	-0.100	-0.057	-0.026	-0.008	0
	0.15	-0.253	-0.218	-0.124	-0.090	-0.076	-0.065	-0.065	-0.041	-0.023	-0.010	0
	0.20	-0.143	-0.121	-0.074	-0.062	-0.058	-0.056	-0.056	-0.037	-0.023	-0.011	0
	0.25	-0.078	-0.068	-0.053	-0.052	-0.052	-0.053	-0.053	-0.037	-0.023	-0.011	0
	0.30	-0.045	-0.043	-0.045	-0.049	-0.050	-0.052	-0.052	-0.037	-0.023	-0.011	0
	0.40	-0.029	-0.033	-0.044	-0.049	-0.051	-0.052	-0.052	-0.037	-0.023	-0.011	0
	0.50	-0.036	-0.039	-0.048	-0.051	-0.052	-0.053	-0.053	-0.037	-0.023	-0.011	0
	0.60	-0.044	-0.047	-0.051	-0.052	-0.053	-0.054	-0.054	-0.037	-0.023	-0.010	0
	0.70	-0.050	-0.051	-0.053	-0.053	-0.054	-0.054	-0.054	-0.037	-0.023	-0.010	0
	0.80	-0.054	-0.054	-0.054	-0.054	-0.054	-0.054	-0.054	-0.037	-0.023	-0.010	0
$\bar{\tau}_{rz} / \eta_1$	0	0	0.149	0.335	0.432	0.511	+∞	+∞	0.527	0.468	0.430	0
	0.01	0	0.148	0.331	0.422	0.488	0.555	0.555	0.493	0.427	0.316	0
	0.02	0	0.145	0.319	0.396	0.437	0.457	0.457	0.419	0.336	0.179	0
	0.05	0	0.126	0.248	0.271	0.266	0.242	0.242	0.192	0.116	0.037	0
	0.10	0	0.074	0.102	0.081	0.062	0.039	0.039	0.014	-0.006	-0.014	0
	0.15	0	0.022	0.002	-0.018	-0.028	-0.036	-0.036	-0.040	-0.037	-0.024	0
	0.20	0	-0.014	-0.045	-0.055	-0.057	-0.056	-0.056	-0.051	-0.041	-0.024	0
	0.25	0	-0.031	-0.060	-0.062	-0.060	-0.055	-0.055	-0.047	-0.036	-0.020	0
	0.30	0	-0.035	-0.057	-0.055	-0.052	-0.046	-0.046	-0.038	-0.028	-0.016	0
	0.40	0	-0.026	-0.036	-0.033	-0.030	-0.026	-0.026	-0.020	-0.014	-0.008	0
	0.50	0	-0.014	-0.018	-0.015	-0.014	-0.011	-0.011	-0.009	-0.006	-0.003	0
	0.60	0	-0.006	-0.007	-0.006	-0.005	-0.004	-0.004	-0.003	-0.002	-0.001	0
	0.70	0	-0.002	-0.002	-0.002	-0.002	-0.001	-0.001	-0.001	-0.001	0.000	0
	0.80	0	0.000	0.000	0.000	0.000	0.000	0.000	0.000	0.000	0.000	0

	λ / p	Core							Shell			
		0	0.3	0.6	0.7	0.75	0.80	0.85	0.85	0.90	0.95	1.00
$\dfrac{\bar{\sigma}_z}{\eta_1}$	0	0	0	0	0	0	0	−0.15	0.15	0	0	0.6
	0.01	−0.030	−0.031	−0.039	−0.046	−0.055	−0.078	−0.169	0.126	0.044	0.068	0.656
	0.02	−0.059	−0.062	−0.075	−0.088	−0.101	−0.127	−0.176	0.122	0.099	0.210	0.696
	0.05	−0.141	−0.147	−0.163	−0.167	−0.165	−0.152	−0.113	0.187	0.280	0.465	0.745
	0.10	−0.246	−0.247	−0.218	−0.173	−0.134	−0.077	0.003	0.303	0.409	0.540	0.689
	0.15	−0.299	−0.286	−0.203	−0.134	−0.087	−0.029	0.039	0.339	0.416	0.500	0.587
	0.20	−0.304	−0.279	−0.171	−0.103	−0.062	−0.017	0.032	0.333	0.384	0.437	0.488
	0.25	−0.279	−0.249	−0.142	−0.086	−0.055	−0.022	0.011	0.311	0.345	0.377	0.406
	0.30	−0.241	−0.211	−0.120	−0.078	−0.056	−0.034	−0.012	0.288	0.309	0.327	0.343
	0.40	−0.167	−0.147	−0.096	−0.076	−0.066	−0.057	−0.049	0.251	0.257	0.263	0.266
	0.50	−0.119	−0.109	−0.086	−0.078	−0.075	−0.072	−0.070	0.230	0.232	0.232	0.232
	0.60	−0.096	−0.092	−0.083	−0.081	−0.080	−0.079	−0.079	0.221	0.221	0.220	0.219
	0.70	−0.086	−0.085	−0.083	−0.082	−0.082	−0.082	−0.083	0.217	0.217	0.217	0.216
	0.80	−0.083	−0.083	−0.083	−0.083	−0.083	−0.083	−0.083	0.217	0.217	0.217	0.217
$\dfrac{\bar{\sigma}_\theta}{\eta_1}$	0	−0.9	−0.9	−0.9	−0.9	−0.9	−0.9	−0.9	−0.6	−0.6	−0.6	−0.4
	0.01	−0.846	−0.843	−0.831	−0.822	−0.815	−0.805	−0.788	−0.488	−0.473	−0.429	−0.278
	0.02	−0.793	−0.787	−0.763	−0.745	−0.732	−0.714	−0.686	−0.386	−0.359	−0.301	−0.194
	0.05	−0.637	−0.624	−0.572	−0.536	−0.512	−0.482	−0.443	−0.143	−0.111	−0.068	−0.016
	0.10	−0.411	−0.391	−0.323	−0.284	−0.261	−0.235	−0.207	0.093	0.106	0.123	0.142
	0.15	−0.241	−0.223	−0.166	−0.138	−0.123	−0.107	−0.091	0.209	0.209	0.211	0.214
	0.20	−0.128	−0.115	−0.078	−0.062	−0.054	−0.045	−0.037	0.263	0.255	0.249	0.244
	0.25	−0.063	−0.055	−0.035	−0.027	−0.023	−0.019	−0.015	0.285	0.272	0.262	0.253
	0.30	−0.030	−0.026	−0.017	−0.014	−0.013	−0.011	−0.010	0.290	0.275	0.262	0.251
	0.40	−0.015	−0.015	−0.016	−0.016	−0.017	−0.017	−0.018	0.282	0.265	0.251	0.239
	0.50	−0.023	−0.023	−0.026	−0.027	−0.028	−0.028	−0.029	0.271	0.254	0.240	0.228
	0.60	−0.032	−0.032	−0.034	−0.035	−0.035	−0.036	−0.036	0.264	0.247	0.233	0.221
	0.70	−0.037	−0.038	−0.039	−0.039	−0.039	−0.039	−0.040	0.260	0.244	0.230	0.218
	0.80	−0.041	−0.041	−0.041	−0.041	−0.041	−0.041	−0.041	0.258	0.242	0.229	0.217
$\dfrac{\bar{\sigma}_r}{\eta_1}$	0	−0.9	−0.9	−0.9	−0.9	−0.9	−0.9	−0.75	−0.75	−0.6	−0.6	0
	0.01	−0.846	−0.840	−0.808	−0.779	−0.754	−0.705	−0.576	−0.576	−0.400	−0.212	0
	0.02	−0.793	−0.780	−0.718	−0.664	−0.619	−0.545	−0.421	−0.421	−0.249	−0.074	0
	0.05	−0.637	−0.608	−0.481	−0.389	−0.326	−0.250	−0.166	−0.166	−0.073	−0.015	0
	0.10	−0.411	−0.369	−0.228	−0.158	−0.122	−0.090	−0.063	−0.063	−0.030	−0.010	0
	0.15	−0.241	−0.206	−0.111	−0.077	−0.063	−0.052	−0.046	−0.046	−0.026	−0.011	0
	0.20	−0.128	−0.107	−0.060	−0.049	−0.045	−0.043	−0.042	−0.042	−0.026	−0.012	0
	0.25	−0.063	−0.054	−0.040	−0.039	−0.039	−0.040	−0.041	−0.041	−0.026	−0.013	0
	0.30	−0.030	−0.029	−0.032	−0.036	−0.038	−0.040	−0.041	−0.041	−0.026	−0.013	0
	0.40	−0.015	−0.019	−0.031	−0.036	−0.038	−0.040	−0.041	−0.041	−0.026	−0.012	0
	0.50	−0.023	−0.026	−0.035	−0.038	−0.040	−0.041	−0.041	−0.041	−0.026	−0.012	0
	0.60	−0.032	−0.034	−0.039	−0.040	−0.041	−0.041	−0.041	−0.041	−0.026	−0.012	0
	0.70	−0.037	−0.038	−0.040	−0.041	−0.041	−0.041	−0.041	−0.041	−0.025	−0.012	0
	0.80	−0.041	−0.041	−0.041	−0.041	−0.041	−0.041	−0.041	−0.041	−0.025	−0.012	0
$\dfrac{\bar{\tau}_{rz}}{\eta_1}$	0	0	0.145	0.321	0.400	0.452	0.529	$+\infty$	$+\infty$	0.532	0.458	0
	0.01	0	0.144	0.317	0.393	0.440	0.502	0.558	0.558	0.480	0.341	0
	0.02	0	0.142	0.306	0.372	0.408	0.442	0.445	0.445	0.370	0.199	0
	0.05	0	0.123	0.239	0.260	0.258	0.239	0.195	0.195	0.122	0.041	0
	0.10	0	0.071	0.096	0.075	0.057	0.034	0.010	0.010	−0.010	−0.016	0
	0.15	0	0.019	−0.004	−0.024	−0.034	−0.041	−0.045	−0.045	−0.041	−0.027	0
	0.20	0	−0.017	−0.051	−0.060	−0.062	−0.061	−0.055	−0.055	−0.043	−0.025	0
	0.25	0	−0.034	−0.064	−0.066	−0.064	−0.059	−0.050	−0.050	−0.038	−0.021	0
	0.30	0	−0.037	−0.060	−0.058	−0.054	−0.048	−0.040	−0.040	−0.029	−0.016	0
	0.40	0	−0.027	−0.037	−0.034	−0.031	−0.026	−0.021	−0.021	−0.015	−0.008	0
	0.50	0	−0.014	−0.018	−0.016	−0.014	−0.012	−0.009	−0.009	−0.006	−0.003	0
	0.60	0	−0.006	−0.007	−0.006	−0.005	−0.004	−0.003	−0.003	−0.002	−0.001	0
	0.70	0	−0.002	−0.002	−0.002	−0.002	−0.001	−0.001	−0.001	−0.001	0.000	0
	0.80	0	0.000	0.000	0.000	0.000	0.000	0.000	0.000	0.000	0.000	0

$\gamma = 1,5 \, ; \, \mu = 0,9 \, ; \, \delta = 1$

	λ / p	Core								Shell		
		0	0.3	0.6	0.7	0.75	0.80	0.85	**0.90**	0.90	**0.95**	1.00
$\dfrac{\bar{\sigma}_z}{\eta_1}$	0	0	0	0	0	0	0	0	−0.15	0.15	0	0.6
	0.01	−0.029	−0.031	−0.036	−0.040	−0.044	−0.052	−0.072	−0.158	0.146	0.102	0.694
	0.02	−0.058	−0.061	−0.071	−0.077	−0.083	−0.092	−0.108	−0.128	0.171	0.261	0.760
	0.05	−0.139	−0.144	−0.154	−0.150	−0.140	−0.119	−0.071	0.031	0.330	0.525	0.824
	0.10	−0.242	−0.240	−0.203	−0.151	−0.107	−0.045	0.041	0.154	0.454	0.593	0.750
	0.15	−0.291	−0.275	−0.184	−0.109	−0.058	0.003	0.075	0.156	0.456	0.544	0.633
	0.20	−0.292	−0.265	−0.149	−0.077	−0.034	0.014	0.066	0.120	0.419	0.474	0.526
	0.25	−0.263	−0.231	−0.118	−0.059	−0.026	0.008	0.042	0.077	0.377	0.409	0.439
	0.30	−0.222	−0.191	−0.095	−0.049	−0.028	−0.005	0.017	0.038	0.338	0.357	0.373
	0.40	−0.144	−0.123	−0.070	−0.051	−0.039	−0.030	−0.022	−0.015	0.285	0.290	0.293
	0.50	−0.094	−0.084	−0.060	−0.053	−0.049	−0.046	−0.043	−0.042	0.258	0.259	0.258
	0.60	−0.070	−0.066	−0.055	−0.055	−0.054	−0.053	−0.053	−0.053	0.247	0.246	0.246
	0.70	−0.060	−0.059	−0.056	−0.056	−0.056	−0.056	−0.056	−0.057	0.243	0.243	0.242
	0.80	−0.057	−0.057	−0.057	−0.057	−0.057	−0.057	−0.057	−0.057	0.243	0.243	0.243
$\dfrac{\bar{\sigma}_\theta}{\eta_1}$	0	−0.9	−0.9	−0.9	−0.9	−0.9	−0.9	−0.9	−0.9	−0.6	−0.6	−0.4
	0.01	−0.845	−0.842	−0.829	−0.820	−0.813	−0.804	−0.789	−0.763	−0.463	−0.421	−0.266
	0.02	−0.790	−0.784	−0.760	−0.742	−0.729	−0.711	−0.685	−0.642	−0.342	−0.284	−0.174
	0.05	−0.631	−0.617	−0.564	−0.528	−0.503	−0.472	−0.433	−0.383	−0.083	−0.039	0.014
	0.10	−0.399	−0.379	−0.309	−0.269	−0.245	−0.219	−0.190	−0.160	0.140	0.156	0.174
	0.15	−0.226	−0.207	−0.149	−0.121	−0.106	−0.090	−0.073	−0.057	0.243	0.244	0.246
	0.20	−0.112	−0.099	−0.061	−0.045	−0.036	−0.028	−0.019	−0.011	0.289	0.281	0.275
	0.25	−0.046	−0.038	−0.018	−0.010	−0.006	−0.002	0.001	0.005	0.305	0.293	0.282
	0.30	−0.014	−0.010	−0.001	0.002	0.003	0.005	0.006	0.007	0.307	0.292	0.280
	0.40	0.000	0.000	−0.002	−0.002	−0.002	−0.003	−0.003	−0.004	0.296	0.280	0.266
	0.50	−0.009	−0.009	−0.012	−0.013	−0.014	−0.015	−0.015	−0.016	0.284	0.268	0.254
	0.60	−0.018	−0.019	−0.021	−0.021	−0.022	−0.022	−0.023	−0.023	0.277	0.261	0.247
	0.70	−0.024	−0.025	−0.025	−0.026	−0.026	−0.026	−0.026	−0.027	0.273	0.258	0.244
	0.80	−0.028	−0.028	−0.028	−0.028	−0.028	−0.028	−0.028	−0.028	0.272	0.256	0.243
$\dfrac{\bar{\sigma}_r}{\eta_1}$	0	−0.9	−0.9	−0.9	−0.9	−0.9	−0.9	−0.9	−0.75	−0.75	−0.6	0
	0.01	−0.845	−0.838	−0.807	−0.780	−0.758	−0.723	−0.659	−0.490	−0.490	−0.227	0
	0.02	−0.790	−0.777	−0.715	−0.665	−0.624	−0.564	−0.466	−0.303	−0.303	−0.089	0
	0.05	−0.631	−0.601	−0.474	−0.383	−0.323	−0.249	−0.166	−0.086	−0.086	−0.019	0
	0.10	−0.399	−0.357	−0.215	−0.145	−0.109	−0.076	−0.050	−0.033	−0.033	−0.012	0
	0.15	−0.226	−0.190	−0.095	−0.062	−0.049	−0.038	−0.032	−0.029	−0.029	−0.013	0
	0.20	−0.112	−0.091	−0.047	−0.034	−0.031	−0.029	−0.028	−0.029	−0.029	−0.014	0
	0.25	−0.046	−0.037	−0.025	−0.024	−0.025	−0.026	−0.028	−0.029	−0.029	−0.014	0
	0.30	−0.014	−0.013	−0.018	−0.022	−0.024	−0.026	−0.028	−0.029	−0.029	−0.014	0
	0.40	0.000	−0.005	−0.020	−0.023	−0.025	−0.027	−0.028	−0.028	−0.028	−0.014	0
	0.50	−0.009	−0.013	−0.022	−0.025	−0.026	−0.028	−0.028	−0.028	−0.028	−0.014	0
	0.60	−0.018	−0.021	−0.026	−0.027	−0.028	−0.028	−0.028	−0.028	−0.028	−0.013	0
	0.70	−0.024	−0.025	−0.027	−0.028	−0.028	−0.028	−0.028	−0.028	−0.028	−0.013	0
	0.80	−0.028	−0.028	−0.028	−0.028	−0.028	−0.028	−0.028	−0.028	−0.028	−0.013	0
$\dfrac{\bar{\tau}_{rz}}{\eta_1}$	0	0	0.143	0.311	0.380	0.422	0.471	0.543	$+\infty$	$+\infty$	0.514	0
	0.01	0	0.142	0.307	0.374	0.412	0.455	0.507	0.542	0.542	0.386	0
	0.02	0	0.139	0.296	0.355	0.384	0.412	0.429	0.396	0.396	0.225	0
	0.05	0	0.120	0.230	0.249	0.246	0.228	0.188	0.120	0.120	0.040	0
	0.10	0	0.068	0.088	0.066	0.047	0.024	0.001	−0.017	−0.017	−0.021	0
	0.15	0	0.015	−0.012	−0.032	−0.042	−0.049	−0.052	−0.046	−0.046	−0.030	0
	0.20	0	−0.020	−0.055	−0.067	−0.069	−0.067	−0.060	−0.047	−0.047	−0.027	0
	0.25	0	−0.037	−0.069	−0.071	−0.068	−0.063	−0.053	−0.040	−0.040	−0.022	0
	0.30	0	−0.040	−0.064	−0.062	−0.057	−0.051	−0.042	−0.031	−0.031	−0.017	0
	0.40	0	−0.028	−0.036	−0.035	−0.032	−0.027	−0.022	−0.015	−0.015	−0.008	0
	0.50	0	−0.015	−0.019	−0.016	−0.014	−0.012	−0.009	−0.007	−0.007	−0.003	0
	0.60	0	−0.006	−0.006	−0.006	−0.005	−0.004	−0.003	−0.002	−0.002	−0.001	0
	0.70	0	−0.002	−0.002	−0.002	−0.002	−0.001	−0.001	−0.001	−0.001	0.000	0
	0.80	0	0.000	0.000	0.000	0.000	0.000	0.000	0.000	0.000	0.000	0

	λ	Core									Shell	
	p	**0**	**0.3**	**0.6**	**0.7**	**0.75**	**0.80**	**0.85**	**0.90**	**0.95**	**0.95**	**1.00**
$\bar\sigma_z / \eta_1$	0	0	0	0	0	0	0	0	0	-0.15	0.15	0.6
	0.01	-0.029	-0.030	-0.035	-0.037	-0.038	-0.041	-0.045	-0.054	-0.088	-0.214	0.795
	0.02	-0.058	-0.060	-0.068	-0.072	-0.072	-0.074	-0.071	-0.054	0.053	0.355	0.893
	0.05	-0.138	-0.142	-0.147	-0.137	-0.123	-0.091	-0.031	0.088	0.305	0.606	0.933
	0.10	-0.239	-0.236	-0.189	-0.130	-0.080	-0.011	0.083	0.206	0.354	0.653	0.818
	0.15	-0.284	-0.266	-0.164	-0.083	-0.028	0.038	0.114	0.200	0.291	0.591	0.684
	0.20	-0.281	-0.251	-0.125	-0.049	-0.003	0.048	0.102	0.158	0.213	0.513	0.566
	0.25	-0.247	-0.212	-0.092	-0.030	0.004	0.039	0.075	0.111	0.144	0.444	0.474
	0.30	-0.202	-0.169	-0.068	-0.023	0.001	0.025	0.048	0.069	0.089	0.388	0.404
	0.40	-0.119	-0.097	-0.042	-0.021	-0.011	-0.002	0.007	0.013	0.018	0.318	0.321
	0.50	-0.067	-0.056	-0.032	-0.024	-0.020	-0.018	-0.016	-0.014	-0.014	0.286	0.286
	0.60	-0.042	-0.038	-0.029	-0.027	-0.026	-0.025	-0.025	-0.025	-0.026	0.274	0.273
	0.70	-0.032	-0.031	-0.029	-0.028	-0.028	-0.028	-0.029	-0.029	-0.029	0.270	0.270
	0.80	-0.029	-0.029	-0.029	-0.029	-0.029	-0.029	-0.029	-0.029	-0.029	0.271	0.271
$\bar\sigma_\theta / \eta_1$	0	-0.9	-0.9	-0.9	-0.9	-0.9	-0.9	-0.9	-0.9	-0.9	-0.6	-0.4
	0.01	-0.843	-0.840	-0.827	-0.817	-0.810	-0.800	-0.785	-0.760	-0.704	-0.404	-0.241
	0.02	-0.787	-0.781	-0.755	-0.737	-0.723	-0.704	-0.677	-0.634	-0.557	-0.257	-0.138
	0.05	-0.623	-0.609	-0.553	-0.515	-0.489	-0.457	-0.416	-0.364	-0.301	-0.001	0.055
	0.10	-0.385	-0.365	-0.292	-0.251	-0.226	-0.198	-0.169	-0.137	-0.105	0.195	0.212
	0.15	-0.208	-0.189	-0.130	-0.101	-0.085	-0.069	-0.052	-0.036	-0.019	0.281	0.281
	0.20	-0.093	-0.080	-0.041	-0.025	-0.016	-0.008	0.000	0.008	0.016	0.316	0.308
	0.25	-0.027	-0.020	0.001	0.009	0.012	0.016	0.020	0.023	0.026	0.326	0.314
	0.30	0.004	0.008	0.016	0.019	0.021	0.022	0.023	0.024	0.024	0.324	0.310
	0.40	0.016	0.016	0.015	0.014	0.013	0.012	0.012	0.011	0.010	0.310	0.295
	0.50	0.006	0.005	0.003	0.001	0.001	0.000	-0.001	-0.002	-0.002	0.298	0.282
	0.60	-0.004	-0.005	-0.007	-0.007	-0.008	-0.008	-0.009	-0.009	-0.010	0.290	0.275
	0.70	-0.010	-0.011	-0.012	-0.012	-0.012	-0.012	-0.013	-0.013	-0.013	0.288	0.272
	0.80	-0.014	-0.014	-0.014	-0.014	-0.014	-0.014	-0.014	-0.014	-0.014	0.285	0.271
$\bar\sigma_r / \eta_1$	0	-0.9	-0.9	-0.9	-0.9	-0.9	-0.9	-0.9	-0.9	-0.75	-0.75	0
	0.01	-0.843	-0.836	-0.804	-0.777	-0.756	-0.723	-0.670	-0.567	-0.302	-0.302	0
	0.02	-0.787	-0.773	-0.710	-0.659	-0.620	-0.561	-0.474	-0.332	-0.119	-0.119	0
	0.05	-0.623	-0.592	-0.461	-0.369	-0.307	-0.235	-0.152	-0.073	-0.020	-0.020	0
	0.10	-0.385	-0.342	-0.197	-0.126	-0.091	-0.058	-0.033	-0.018	-0.013	-0.013	0
	0.15	-0.208	-0.172	-0.077	-0.045	-0.031	-0.022	-0.016	-0.014	-0.014	-0.014	0
	0.20	-0.093	-0.072	-0.028	-0.018	-0.015	-0.014	-0.014	-0.015	-0.015	-0.015	0
	0.25	-0.027	-0.019	-0.008	-0.009	-0.011	-0.012	-0.014	-0.015	-0.016	-0.016	0
	0.30	0.004	0.004	-0.002	-0.007	-0.009	-0.012	-0.014	-0.015	-0.016	-0.016	0
	0.40	0.016	0.011	-0.003	-0.008	-0.011	-0.013	-0.014	-0.015	-0.015	-0.015	0
	0.50	0.006	0.002	-0.008	-0.011	-0.013	-0.014	-0.014	-0.014	-0.015	-0.015	0
	0.60	-0.004	-0.006	-0.011	-0.013	-0.014	-0.014	-0.014	-0.014	-0.014	-0.014	0
	0.70	-0.010	-0.011	-0.013	-0.014	-0.014	-0.014	-0.014	-0.014	-0.014	-0.014	0
	0.80	-0.014	-0.014	-0.014	-0.014	-0.014	-0.014	-0.014	-0.014	-0.014	-0.014	0
$\bar\tau_{rz} / \eta_1$	0	0	0.142	0.306	0.370	0.404	0.444	0.489	0.549	$+\infty$	$+\infty$	0
	0.01	0	0.141	0.302	0.363	0.395	0.430	0.464	0.492	0.444	0.444	0
	0.02	0	0.138	0.291	0.344	0.370	0.390	0.400	0.374	0.234	0.234	0
	0.05	0	0.119	0.223	0.238	0.233	0.213	0.171	0.104	0.028	0.028	0
	0.10	0	0.064	0.078	0.054	0.034	0.010	-0.013	-0.029	-0.028	-0.028	0
	0.15	0	0.011	-0.021	-0.043	-0.053	-0.060	-0.061	-0.053	-0.033	-0.033	0
	0.20	0	-0.025	-0.065	-0.075	-0.076	-0.074	-0.066	-0.051	-0.029	-0.029	0
	0.25	0	-0.041	-0.075	-0.076	-0.074	-0.067	-0.057	-0.042	-0.023	-0.023	0
	0.30	0	-0.043	-0.068	-0.065	-0.061	-0.054	-0.044	-0.032	-0.017	-0.017	0
	0.40	0	-0.030	-0.041	-0.037	-0.033	-0.028	-0.023	-0.016	-0.008	-0.008	0
	0.50	0	-0.015	-0.019	-0.017	-0.015	-0.012	-0.010	-0.007	-0.003	-0.003	0
	0.60	0	-0.006	-0.007	-0.006	-0.005	-0.004	-0.003	-0.002	-0.001	-0.001	0
	0.70	0	-0.002	-0.002	-0.002	-0.001	-0.001	-0.001	-0.001	0.000	0.000	0
	0.80	0	0.000	0.000	0.000	0.000	0.000	0.000	0.000	0.000	0.000	0

$\gamma = 2;\ \mu = 0{,}7;\ \delta = 1$

	λ / p	Core				Shell						
		0	0.3	0.6	0.7	0.7	0.75	0.80	0.85	0.90	0.95	1.00
$\dfrac{\bar{\sigma}_z}{\eta_1}$	0	0	0	0	−0.30	0.30	0	0	0	0	0	0.6
	0.01	−0.044	−0.049	−0.094	−0.337	0.260	0.073	0.029	0.016	0.018	0.057	0.640
	0.02	−0.088	−0.097	−0.177	−0.370	0.233	0.110	0.057	0.043	0.065	0.191	0.668
	0.05	−0.212	−0.230	−0.337	−0.420	0.179	0.147	0.140	0.174	0.269	0.452	0.720
	0.10	−0.376	−0.392	−0.426	−0.410	0.190	0.217	0.265	0.338	0.439	0.568	0.718
	0.15	−0.471	−0.470	−0.423	−0.367	0.233	0.276	0.330	0.396	0.473	0.560	0.653
	0.20	−0.504	−0.487	−0.397	−0.333	0.267	0.307	0.352	0.402	0.457	0.514	0.571
	0.25	−0.495	−0.469	−0.369	−0.313	0.286	0.318	0.352	0.388	0.425	0.460	0.495
	0.30	−0.466	−0.438	−0.347	−0.304	0.296	0.320	0.344	0.368	0.391	0.413	0.432
	0.40	−0.396	−0.376	−0.321	−0.299	0.301	0.312	0.322	0.331	0.340	0.346	0.351
	0.50	−0.347	−0.336	−0.310	−0.301	0.299	0.303	0.307	0.310	0.312	0.313	0.313
	0.60	−0.321	−0.317	−0.306	−0.303	0.297	0.298	0.299	0.299	0.299	0.299	0.298
	0.70	−0.310	−0.309	−0.306	−0.305	0.295	0.295	0.295	0.295	0.295	0.294	0.294
	0.80	−0.306	−0.306	−0.306	−0.306	0.294	0.294	0.294	0.294	0.294	0.294	0.294
$\dfrac{\bar{\sigma}_\theta}{\eta_1}$	0	−1.2	−1.2	−1.2	−1.2	−0.6	−0.6	−0.6	−0.6	−0.6	−0.6	−0.4
	0.01	−1.136	−1.133	−1.116	−1.093	−0.493	−0.502	−0.500	−0.492	−0.474	−0.430	−0.281
	0.02	−1.073	−1.066	−1.033	−0.999	−0.399	−0.408	−0.404	−0.389	−0.359	−0.301	−0.196
	0.05	−0.890	−0.873	−0.805	−0.756	−0.156	−0.162	−0.155	−0.135	−0.104	−0.062	−0.014
	0.10	−0.622	−0.598	−0.514	−0.468	0.132	0.121	0.119	0.125	0.136	0.151	0.158
	0.15	−0.419	−0.397	−0.329	−0.296	0.304	0.283	0.270	0.262	0.259	0.259	0.261
	0.20	−0.283	−0.267	−0.222	−0.202	0.398	0.370	0.349	0.333	0.321	0.312	0.305
	0.25	−0.201	−0.191	−0.165	−0.154	0.446	0.412	0.386	0.365	0.348	0.334	0.322
	0.30	−0.156	−0.151	−0.138	−0.134	0.466	0.430	0.401	0.376	0.356	0.340	0.325
	0.40	−0.130	−0.129	−0.129	−0.129	0.471	0.432	0.401	0.374	0.352	0.333	0.316
	0.50	−0.134	−0.135	−0.137	−0.138	0.462	0.423	0.391	0.364	0.342	0.322	0.306
	0.60	−0.142	−0.143	−0.145	−0.145	0.455	0.416	0.383	0.357	0.334	0.315	0.299
	0.70	−0.148	−0.149	−0.150	−0.150	0.450	0.411	0.379	0.353	0.331	0.312	0.296
	0.80	−0.152	−0.152	−0.152	−0.152	0.447	0.409	0.377	0.351	0.329	0.310	0.294
$\dfrac{\bar{\sigma}_r}{\eta_1}$	0	−1.2	−1.2	−1.2	−0.90	−0.90	−0.6	−0.6	−0.6	−0.6	−0.6	0
	0.01	−1.136	−1.127	−1.063	−0.813	−0.813	−0.591	−0.515	−0.456	−0.375	−0.208	0
	0.02	−1.073	−1.055	−0.937	−0.720	−0.720	−0.547	−0.433	−0.335	−0.217	−0.066	0
	0.05	−0.890	−0.850	−0.650	−0.494	−0.494	−0.366	−0.249	−0.146	−0.062	−0.011	0
	0.10	−0.622	−0.568	−0.376	−0.285	−0.285	−0.202	−0.132	−0.076	−0.037	−0.012	0
	0.15	−0.419	−0.373	−0.247	−0.203	−0.203	−0.145	−0.099	−0.062	−0.035	−0.015	0
	0.20	−0.283	−0.254	−0.187	−0.169	−0.169	−0.124	−0.087	−0.058	−0.035	−0.016	0
	0.25	−0.201	−0.186	−0.160	−0.155	−0.155	−0.115	−0.083	−0.057	−0.035	−0.017	0
	0.30	−0.156	−0.152	−0.148	−0.150	−0.150	−0.112	−0.081	−0.056	−0.035	−0.017	0
	0.40	−0.130	−0.133	−0.144	−0.148	−0.148	−0.111	−0.081	−0.056	−0.035	−0.017	0
	0.50	−0.134	−0.138	−0.147	−0.150	−0.150	−0.112	−0.082	−0.056	−0.035	−0.016	0
	0.60	−0.142	−0.145	−0.150	−0.151	−0.151	−0.113	−0.082	−0.056	−0.035	−0.016	0
	0.70	−0.148	−0.149	−0.152	−0.152	−0.152	−0.114	−0.083	−0.056	−0.035	−0.016	0
	0.80	−0.152	−0.152	−0.152	−0.152	−0.152	−0.114	−0.083	−0.056	−0.035	−0.016	0
$\dfrac{\bar{\tau}_{rz}}{\eta_1}$	0	0	0.222	0.573	$+\infty$	$+\infty$	0.716	0.593	0.526	0.479	0.437	0
	0.01	0	0.220	0.561	0.802	0.802	0.678	0.574	0.505	0.440	0.323	0
	0.02	0	0.216	0.529	0.653	0.653	0.602	0.524	0.449	0.355	0.189	0
	0.05	0	0.190	0.388	0.402	0.402	0.372	0.318	0.241	0.147	0.052	0
	0.10	0	0.119	0.174	0.145	0.145	0.118	0.085	0.049	0.017	−0.002	0
	0.15	0	0.050	0.040	0.015	0.015	0.001	−0.012	−0.022	−0.025	−0.018	0
	0.20	0	0.002	−0.027	−0.040	−0.040	−0.045	−0.046	−0.044	−0.036	−0.022	0
	0.25	0	−0.024	−0.053	−0.057	−0.057	−0.056	−0.053	−0.046	−0.035	−0.020	0
	0.30	0	−0.033	−0.056	−0.055	−0.055	−0.052	−0.047	−0.039	−0.029	−0.016	0
	0.40	0	−0.027	−0.039	−0.036	−0.036	−0.032	−0.028	−0.023	−0.016	−0.008	0
	0.50	0	−0.015	−0.020	−0.017	−0.017	−0.016	−0.013	−0.010	−0.007	−0.004	0
	0.60	0	−0.007	−0.008	−0.007	−0.007	−0.006	−0.005	−0.004	−0.003	−0.001	0
	0.70	0	−0.003	−0.003	−0.002	−0.002	−0.002	−0.002	−0.001	−0.001	0.000	0
	0.80	0	0.000	0.000	0.000	0.000	0.000	0.000	0.000	0.000	0.000	0

	λ	Core					Shell					
	p	0	0.3	0.6	0.7	0.75	0.75	0.80	0.85	0.90	0.95	1.00
$\bar{\sigma}_z/\eta_1$	0	0	0	0	0	−0.30	0.30	0	0	0	0	0.6
	0.01	−0.042	−0.046	−0.073	−0.142	−0.336	0.266	0.078	0.037	0.031	0.070	0.656
	0.02	−0.084	−0.091	−0.141	−0.236	−0.362	0.242	0.123	0.079	0.091	0.216	0.699
	0.05	−0.203	−0.218	−0.292	−0.354	−0.387	0.215	0.198	0.224	0.316	0.501	0.781
	0.10	−0.360	−0.372	−0.387	−0.365	−0.335	0.265	0.315	0.390	0.496	0.632	0.791
	0.15	−0.450	−0.445	−0.386	−0.324	−0.279	0.321	0.379	0.449	0.530	0.622	0.721
	0.20	−0.478	−0.458	−0.358	−0.290	−0.248	0.352	0.401	0.454	0.511	0.572	0.632
	0.25	−0.466	−0.437	−0.329	−0.270	−0.236	0.364	0.400	0.438	0.476	0.514	0.550
	0.30	−0.433	−0.402	−0.306	−0.260	−0.235	0.365	0.390	0.416	0.440	0.463	0.483
	0.40	−0.358	−0.336	−0.278	−0.255	−0.244	0.356	0.367	0.377	0.385	0.392	0.397
	0.50	−0.305	−0.294	−0.267	−0.257	−0.253	0.347	0.351	0.354	0.356	0.357	0.357
	0.60	−0.278	−0.274	−0.263	−0.260	−0.259	0.341	0.342	0.343	0.343	0.343	0.342
	0.70	−0.267	−0.265	−0.262	−0.261	−0.261	0.339	0.339	0.339	0.338	0.338	0.337
	0.80	−0.262	−0.262	−0.262	−0.262	−0.262	0.337	0.337	0.337	0.337	0.337	0.337
$\bar{\sigma}_\theta/\eta_1$	0	−1.2	−1.2	−1.2	−1.2	−1.2	−0.6	−0.6	−0.6	−0.6	−0.6	−0.4
	0.01	−1.135	−1.132	−1.116	−1.102	−1.086	−0.486	−0.492	−0.485	−0.468	−0.424	−0.274
	0.02	−1.071	−1.064	−1.033	−1.007	−0.985	−0.385	−0.388	−0.376	−0.347	−0.289	−0.183
	0.05	−0.884	−0.868	−0.803	−0.756	−0.725	−0.125	−0.123	−0.107	−0.077	−0.035	0.015
	0.10	−0.610	−0.586	−0.503	−0.456	−0.428	0.172	0.166	0.169	−0.178	0.192	0.209
	0.15	−0.402	−0.380	−0.311	−0.277	−0.259	0.341	0.324	0.314	0.308	0.306	0.307
	0.20	−0.262	−0.246	−0.199	−0.179	−0.169	0.431	0.406	0.386	0.372	0.360	0.352
	0.25	−0.178	−0.168	−0.141	−0.130	−0.125	0.475	0.444	0.419	0.399	0.383	0.369
	0.30	−0.132	−0.127	−0.114	−0.109	−0.107	0.493	0.459	0.431	0.408	0.388	0.371
	0.40	−0.106	−0.105	−0.105	−0.105	−0.105	0.495	0.458	0.428	0.402	0.380	0.361
	0.50	−0.111	−0.112	−0.114	−0.115	−0.116	0.484	0.448	0.417	0.391	0.369	0.350
	0.60	−0.120	−0.121	−0.122	−0.123	−0.124	0.476	0.440	0.409	0.383	0.361	0.343
	0.70	−0.126	−0.127	−0.128	−0.128	−0.128	0.472	0.435	0.405	0.379	0.358	0.339
	0.80	−0.131	−0.131	−0.131	−0.131	−0.131	0.469	0.433	0.402	0.378	0.356	0.337
$\bar{\sigma}_r/\eta_1$	0	−1.2	−1.2	−1.2	−1.2	−0.90	−0.90	−0.6	−0.6	−0.6	−0.6	0
	0.01	−1.135	−1.127	−1.077	−0.992	−0.790	−0.790	−0.563	−0.472	−0.381	−0.209	0
	0.02	−1.071	−1.054	−0.960	−0.831	−0.681	−0.681	−0.496	−0.361	−0.227	−0.069	0
	0.05	−0.884	−0.846	−0.666	−0.522	−0.428	−0.428	−0.294	−0.173	−0.075	−0.015	0
	0.10	−0.610	−0.557	−0.371	−0.278	−0.232	−0.232	−0.153	−0.091	−0.045	−0.015	0
	0.15	−0.402	−0.357	−0.231	−0.185	−0.165	−0.165	−0.113	−0.072	−0.041	−0.017	0
	0.20	−0.262	−0.233	−0.167	−0.148	−0.141	−0.141	−0.100	−0.067	−0.040	−0.019	0
	0.25	−0.178	−0.163	−0.138	−0.133	−0.132	−0.132	−0.095	−0.065	−0.040	−0.019	0
	0.30	−0.132	−0.128	−0.125	−0.127	−0.128	−0.128	−0.094	−0.065	−0.040	−0.019	0
	0.40	−0.106	−0.110	−0.121	−0.126	−0.128	−0.128	−0.093	−0.065	−0.040	−0.019	0
	0.50	−0.111	−0.115	−0.124	−0.128	−0.129	−0.129	−0.094	−0.065	−0.040	−0.019	0
	0.60	−0.120	−0.122	−0.128	−0.129	−0.130	−0.130	−0.094	−0.065	−0.040	−0.018	0
	0.70	−0.126	−0.127	−0.130	−0.131	−0.131	−0.131	−0.095	−0.065	−0.040	−0.018	0
	0.80	−0.131	−0.131	−0.131	−0.131	−0.131	−0.131	−0.095	−0.065	−0.040	−0.018	0
$\bar{\tau}_{rz}/\eta_1$	0	0	0.211	0.509	0.735	+∞	+∞	0.731	0.602	0.524	0.460	0
	0.01	0	0.210	0.502	0.698	0.811	0.811	0.689	0.575	0.483	0.345	0
	0.02	0	0.206	0.481	0.623	0.664	0.664	0.604	0.508	0.393	0.209	0
	0.05	0	0.181	0.370	0.403	0.388	0.388	0.343	0.267	0.166	0.062	0
	0.10	0	0.114	0.169	0.144	0.120	0.120	0.087	0.051	0.019	−0.001	0
	0.15	0	0.046	0.035	0.011	−0.003	−0.003	−0.016	−0.025	−0.028	−0.020	0
	0.20	0	−0.002	−0.033	−0.048	−0.050	−0.050	−0.051	−0.048	−0.039	−0.024	0
	0.25	0	−0.027	−0.058	−0.062	−0.061	−0.061	−0.057	−0.049	−0.038	−0.021	0
	0.30	0	−0.036	−0.060	−0.059	−0.056	−0.056	−0.050	−0.042	−0.031	−0.017	0
	0.40	0	−0.029	−0.041	−0.038	−0.034	−0.034	−0.030	−0.024	−0.017	−0.009	0
	0.50	0	−0.016	−0.021	−0.018	−0.016	−0.016	−0.014	−0.011	−0.008	−0.004	0
	0.60	0	−0.007	−0.009	−0.007	−0.006	−0.006	−0.005	−0.004	−0.003	−0.001	0
	0.70	0	−0.003	−0.003	−0.002	−0.002	−0.002	−0.002	−0.001	−0.001	0.000	0
	0.80	0	−0.000	0.000	0.000	0.000	0.000	0.000	0.000	0.000	0.000	0

$\gamma = 2; \mu = 0{,}8; \delta = 1$

	λ	Core						Shell				
	p	0	0.3	0.6	0.7	0.75	0.80	0.80	0.85	0.90	0.95	1.00
$\bar{\sigma}_z$ η_1	0	0	0	0	0	0	-0.30	0.30	0	0	0	0.6
	0.01	-0.041	-0.044	-0.061	-0.089	-0.137	-0.330	0.273	0.088	0.054	0.088	0.680
	0.02	-0.081	-0.087	-0.119	-0.165	-0.226	-0.347	0.256	0.148	0.131	0.251	0.743
	0.05	-0.196	-0.207	-0.257	-0.295	-0.318	-0.332	0.270	0.290	0.378	0.567	0.864
	0.10	-0.346	-0.353	-0.351	-0.319	-0.285	-0.230	0.369	0.451	0.563	0.708	0.879
	0.15	-0.430	-0.421	-0.348	-0.279	-0.229	-0.168	0.433	0.508	0.595	0.693	0.798
	0.20	-0.453	-0.429	-0.318	-0.244	-0.198	-0.146	0.453	0.510	0.572	0.635	0.699
	0.25	-0.436	-0.403	-0.286	-0.223	-0.187	-0.148	0.452	0.492	0.532	0.572	0.609
	0.30	-0.398	-0.365	-0.262	-0.212	-0.186	-0.159	0.441	0.467	0.493	0.517	0.538
	0.40	-0.317	-0.294	-0.232	-0.208	-0.196	-0.185	0.415	0.425	0.434	0.441	0.446
	0.50	-0.261	-0.249	-0.220	-0.210	-0.206	-0.202	0.398	0.401	0.403	0.404	0.404
	0.60	-0.232	-0.227	-0.216	-0.213	-0.212	-0.211	0.389	0.389	0.390	0.389	0.388
	0.70	-0.220	-0.219	-0.216	-0.215	-0.215	-0.214	0.385	0.385	0.385	0.384	0.384
	0.80	-0.216	-0.216	-0.216	-0.216	-0.216	-0.216	0.384	0.384	0.384	0.384	0.384
$\bar{\sigma}_\theta$ η_1	0	-1.2	-1.2	-1.2	-1.2	-1.2	-1.2	-0.6	-0.6	-0.6	-0.6	-0.4
	0.01	-1.134	-1.130	-1.115	-1.103	-1.093	-1.075	-0.475	-0.475	-0.459	-0.416	-0.264
	0.02	-1.068	-1.061	-1.031	-1.008	-0.990	-0.963	-0.363	-0.358	-0.331	-0.273	-0.166
	0.05	-0.876	-0.860	-0.796	-0.751	-0.721	-0.683	-0.083	-0.071	-0.043	-0.001	0.050
	0.10	-0.594	-0.571	-0.487	-0.439	-0.411	-0.379	0.221	0.220	0.228	0.241	0.257
	0.15	-0.381	-0.359	-0.288	-0.254	-0.235	-0.215	0.385	0.371	0.363	0.359	0.358
	0.20	-0.238	-0.221	-0.174	-0.153	-0.142	-0.131	0.469	0.446	0.428	0.414	0.403
	0.25	-0.151	-0.141	-0.114	-0.103	-0.098	-0.092	0.508	0.479	0.455	0.436	0.419
	0.30	-0.106	-0.100	-0.087	-0.083	-0.080	-0.078	0.522	0.490	0.463	0.440	0.421
	0.40	-0.080	-0.080	-0.080	-0.080	-0.080	-0.080	0.520	0.485	0.456	0.431	0.409
	0.50	-0.086	-0.087	-0.090	-0.091	-0.091	-0.092	0.508	0.473	0.444	0.419	0.397
	0.60	-0.096	-0.097	-0.099	-0.100	-0.100	-0.101	0.499	0.465	0.436	0.411	0.390
	0.70	-0.103	-0.103	-0.104	-0.105	-0.105	-0.105	0.495	0.460	0.431	0.407	0.386
	0.80	-0.108	-0.108	-0.108	-0.108	-0.108	-0.108	0.492	0.458	0.429	0.405	0.384
$\bar{\sigma}_r$ η_1	0	-1.2	-1.2	-1.2	-1.2	-1.2	-0.90	-0.90	-0.6	-0.6	-0.6	0
	0.01	-1.134	-1.125	-1.082	-1.034	-0.970	-0.759	-0.759	-0.318	-0.394	-0.212	0
	0.02	-1.068	-1.051	-0.969	-0.882	-0.790	-0.625	-0.625	-0.419	-0.248	-0.074	0
	0.05	-0.876	-0.839	-0.672	-0.542	-0.452	-0.348	-0.348	-0.209	-0.093	-0.021	0
	0.10	-0.594	-0.542	-0.360	-0.267	-0.219	-0.176	-0.176	-0.105	-0.053	-0.019	0
	0.15	-0.381	-0.336	-0.210	-0.164	-0.143	-0.127	-0.127	-0.082	-0.046	-0.020	0
	0.20	-0.238	-0.209	-0.143	-0.124	-0.117	-0.113	-0.113	-0.076	-0.046	-0.021	0
	0.25	-0.151	-0.137	-0.113	-0.109	-0.108	-0.108	-0.108	-0.074	-0.046	-0.022	0
	0.30	-0.106	-0.102	-0.101	-0.103	-0.105	-0.106	-0.106	-0.074	-0.046	-0.022	0
	0.40	-0.080	-0.084	-0.097	-0.102	-0.104	-0.106	-0.106	-0.073	-0.046	-0.022	0
	0.50	-0.086	-0.091	-0.101	-0.104	-0.106	-0.107	-0.107	-0.074	-0.045	-0.021	0
	0.60	-0.096	-0.099	-0.104	-0.106	-0.107	-0.107	-0.107	-0.074	-0.045	-0.021	0
	0.70	-0.103	-0.104	-0.107	-0.107	-0.108	-0.108	-0.108	-0.074	-0.045	-0.021	0
	0.80	-0.108	-0.108	-0.108	-0.108	-0.108	-0.108	-0.108	-0.074	-0.045	-0.021	0
$\bar{\tau}_{rz}$ η_1	0	0	0.202	0.467	0.617	0.753	$+\infty$	$+\infty$	0.736	0.594	0.493	0
	0.01	0	0.201	0.461	0.602	0.713	0.829	0.829	0.687	0.548	0.378	0
	0.02	0	0.197	0.445	0.563	0.632	0.661	0.661	0.584	0.446	0.236	0
	0.05	0	0.173	0.351	0.389	0.387	0.356	0.356	0.287	0.185	0.073	0
	0.10	0	0.107	0.160	0.138	0.115	0.084	0.084	0.049	0.017	-0.003	0
	0.15	0	0.040	0.026	0.003	-0.011	-0.023	-0.023	-0.031	-0.033	-0.023	0
	0.20	0	-0.007	-0.041	-0.054	-0.058	-0.058	-0.058	-0.054	-0.044	-0.026	0
	0.25	0	-0.031	-0.065	-0.069	-0.067	-0.062	-0.062	-0.054	-0.041	-0.023	0
	0.30	0	-0.039	-0.065	-0.064	-0.060	-0.054	-0.054	-0.045	-0.033	-0.018	0
	0.40	0	-0.031	-0.044	-0.040	-0.036	-0.032	-0.032	-0.025	-0.018	-0.009	0
	0.50	0	-0.017	-0.022	-0.019	-0.017	-0.014	-0.014	-0.011	-0.008	-0.004	0
	0.60	0	-0.008	-0.009	-0.008	-0.007	-0.006	-0.006	-0.004	-0.003	-0.001	0
	0.70	0	-0.003	-0.003	-0.002	-0.002	-0.002	-0.002	-0.001	-0.001	0.000	0
	0.80	0	0.000	0.000	0.000	0.000	0.000	0.000	0.000	0.000	0.000	0

	λ	Core							Shell			
	p	0	0.3	0.6	0.7	0.75	0.80	0.85	0.85	0.90	0.95	1.00
$\dfrac{\bar{\sigma}_z}{\eta_1}$	0	0	0	0	0	0	0	−0.30	0.30	0	0	0.6
	0.01	−0.040	−0.042	−0.054	−0.068	−0.085	−0.131	−0.313	0.278	0.109	0.119	0.719
	0.02	−0.079	−0.084	−0.106	−0.129	−0.156	−0.210	−0.316	0.281	0.207	0.307	0.815
	0.05	−0.190	−0.199	−0.231	−0.249	−0.257	−0.257	−0.232	0.370	0.459	0.657	0.978
	0.10	−0.335	−0.338	−0.317	−0.274	−0.232	−0.170	−0.081	0.519	0.642	0.798	0.984
	0.15	−0.412	−0.399	−0.310	−0.232	−0.177	−0.109	−0.027	0.572	0.667	0.772	0.883
	0.20	−0.429	−0.400	−0.275	−0.194	−0.145	−0.089	−0.028	0.572	0.637	0.704	0.770
	0.25	−0.405	−0.369	−0.241	−0.172	−0.134	−0.093	−0.050	0.550	0.592	0.634	0.672
	0.30	−0.362	−0.326	−0.214	−0.162	−0.134	−0.106	−0.078	0.522	0.549	0.573	0.595
	0.40	−0.273	−0.248	−0.183	−0.158	−0.145	−0.134	−0.123	0.477	0.486	0.493	0.498
	0.50	−0.213	−0.201	−0.171	−0.161	−0.156	−0.152	−0.149	0.451	0.453	0.454	0.454
	0.60	−0.183	−0.178	−0.167	−0.164	−0.162	−0.161	−0.161	0.439	0.439	0.439	0.438
	0.70	−0.171	−0.169	−0.166	−0.165	−0.165	−0.165	−0.165	0.435	0.434	0.434	0.433
	0.80	−0.166	−0.166	−0.166	−0.166	−0.166	−0.166	−0.166	0.433	0.433	0.433	0.433
$\dfrac{\bar{\sigma}_\theta}{\eta_1}$	0	−1.2	−1.2	−1.2	−1.2	−1.2	−1.2	−1.2	−0.6	−0.6	−0.6	−0.4
	0.01	−1.131	−1.128	−1.113	−1.102	−1.093	−1.080	−1.058	−0.458	−0.448	−0.404	−0.251
	0.02	−1.063	−1.056	−1.026	−1.005	−0.989	−0.966	−0.931	−0.331	−0.309	−0.251	−0.142
	0.05	−0.865	−0.849	−0.784	−0.740	−0.710	−0.673	−0.626	−0.026	0.001	0.043	0.096
	0.10	−0.576	−0.551	−0.465	−0.417	−0.387	−0.355	−0.320	0.280	0.286	0.297	0.313
	0.15	−0.356	−0.333	−0.261	−0.225	−0.206	−0.185	−0.164	0.436	0.425	0.418	0.415
	0.20	−0.209	−0.192	−0.144	−0.122	−0.111	−0.100	−0.089	0.511	0.490	0.473	0.459
	0.25	−0.122	−0.111	−0.084	−0.073	−0.067	−0.062	−0.057	0.543	0.516	0.493	0.474
	0.30	−0.076	−0.071	−0.058	−0.053	−0.051	−0.049	−0.047	0.553	0.522	0.496	0.474
	0.40	−0.052	−0.052	−0.052	−0.053	−0.053	−0.053	−0.054	0.546	0.513	0.485	0.461
	0.50	−0.060	−0.061	−0.064	−0.065	−0.066	−0.066	−0.067	0.533	0.500	0.472	0.447
	0.60	−0.071	−0.071	−0.074	−0.075	−0.075	−0.075	−0.076	0.524	0.491	0.463	0.439
	0.70	−0.073	−0.078	−0.079	−0.080	−0.080	−0.080	−0.081	0.519	0.487	0.459	0.435
	0.80	−0.083	−0.083	−0.083	−0.083	−0.083	−0.083	−0.083	0.517	0.485	0.457	0.433
$\dfrac{\bar{\sigma}_r}{\eta_1}$	0	−1.2	−1.2	−1.2	−1.2	−1.2	−1.2	−0.90	−0.90	−0.6	−0.6	0
	0.01	−1.131	−1.123	−1.083	−1.046	−1.010	−0.937	−0.715	−0.715	−0.435	−0.220	0
	0.02	−1.063	−1.047	−0.969	−0.899	−0.836	−0.729	−0.545	−0.545	−0.298	−0.087	0
	0.05	−0.865	−0.828	−0.667	−0.548	−0.465	−0.363	−0.251	−0.251	−0.118	−0.029	0
	0.10	−0.576	−0.523	−0.341	−0.249	−0.201	−0.156	−0.118	−0.118	−0.060	−0.022	0
	0.15	−0.356	−0.311	−0.185	−0.138	−0.118	−0.102	−0.091	−0.091	−0.052	−0.023	0
	0.20	−0.209	−0.180	−0.115	−0.098	−0.091	−0.087	−0.084	−0.084	−0.051	−0.024	0
	0.25	−0.122	−0.108	−0.085	−0.082	−0.082	−0.082	−0.083	−0.083	−0.052	−0.025	0
	0.30	−0.076	−0.073	−0.074	−0.077	−0.079	−0.081	−0.083	−0.083	−0.052	−0.025	0
	0.40	−0.052	−0.057	−0.071	−0.077	−0.079	−0.081	−0.083	−0.083	−0.051	−0.024	0
	0.50	−0.060	−0.065	−0.076	−0.079	−0.081	−0.082	−0.083	−0.083	−0.051	−0.024	0
	0.60	−0.071	−0.073	−0.079	−0.081	−0.082	−0.083	−0.083	−0.083	−0.051	−0.024	0
	0.70	−0.078	−0.079	−0.082	−0.082	−0.083	−0.083	−0.083	−0.083	−0.051	−0.023	0
	0.80	−0.083	−0.083	−0.083	−0.083	−0.083	−0.083	−0.083	−0.083	−0.051	−0.023	0
$\dfrac{\bar{\tau}_{rz}}{\eta_1}$	0	0	0.196	0.437	0.554	0.635	0.765	$+\infty$	$+\infty$	0.723	0.549	0
	0.01	0	0.194	0.432	0.544	0.618	0.721	0.817	0.817	0.654	0.429	0
	0.02	0	0.191	0.418	0.515	0.572	0.630	0.637	0.637	0.515	0.275	0
	0.05	0	0.167	0.332	0.368	0.371	0.350	0.293	0.293	0.197	0.081	0
	0.10	0	0.101	0.147	0.125	0.103	0.073	0.041	0.041	0.010	−0.008	0
	0.15	0	0.034	0.015	−0.010	−0.023	−0.034	−0.041	−0.041	−0.040	−0.028	0
	0.20	0	−0.013	−0.051	−0.064	−0.068	−0.067	−0.062	−0.062	−0.050	−0.029	0
	0.25	0	−0.037	−0.073	−0.077	−0.075	−0.069	−0.059	−0.059	−0.045	−0.025	0
	0.30	0	−0.044	−0.072	−0.070	−0.066	−0.059	−0.049	−0.049	−0.036	−0.020	0
	0.40	0	−0.033	−0.046	−0.042	−0.039	−0.033	−0.027	−0.027	−0.019	−0.010	0
	0.50	0	−0.018	−0.023	−0.020	−0.018	−0.015	−0.012	−0.012	−0.008	−0.004	0
	0.60	0	−0.008	−0.009	−0.008	−0.007	−0.006	−0.004	−0.004	−0.003	−0.001	0
	0.70	0	−0.003	−0.003	−0.003	−0.002	−0.002	−0.001	−0.001	−0.001	0.000	0
	0.80	0	0.000	0.000	0.000	0.000	0.000	0.000	0.000	0.000	0.000	0

$\gamma = 2; \mu = 0,9; \delta = 1$

	λ	Core								Shell		
	p	0	0.3	0.6	0.7	0.75	0.80	0.85	0.90	0.90	0.95	1.00
$\bar{\sigma}_z$ η_1	0	0	0	0	0	0	0	0	−0.30	0.30	0	0.6
	0.01	−0.039	−0.041	−0.049	−0.056	−0.064	−0.078	−0.119	−0.297	0.312	0.186	0.795
	0.02	−0.077	−0.081	−0.097	−0.108	−0.120	−0.139	−0.179	−0.248	0.351	0.408	0.944
	0.05	−0.186	−0.193	−0.211	−0.214	−0.208	−0.191	−0.146	−0.039	0.558	0.779	1.138
	0.10	−0.326	−0.325	−0.287	−0.230	−0.179	−0.107	−0.005	0.132	0.731	0.904	1.105
	0.15	−0.396	−0.379	−0.272	−0.182	−0.120	−0.045	0.045	0.147	0.746	0.859	0.976
	0.20	−0.405	−0.372	−0.231	−0.142	−0.087	−0.027	0.038	0.108	0.707	0.778	0.847
	0.25	−0.373	−0.333	−0.193	−0.118	−0.077	−0.033	0.012	0.056	0.656	0.699	0.738
	0.30	−0.323	−0.284	−0.164	−0.108	−0.078	−0.048	−0.019	0.009	0.608	0.633	0.655
	0.40	−0.226	−0.200	−0.131	−0.104	−0.091	−0.080	−0.069	−0.060	0.540	0.548	0.552
	0.50	−0.162	−0.149	−0.118	−0.108	−0.103	−0.099	−0.096	−0.094	0.506	0.507	0.507
	0.60	−0.131	−0.126	−0.114	−0.111	−0.110	−0.109	−0.108	−0.108	0.491	0.491	0.490
	0.70	−0.118	−0.116	−0.114	−0.113	−0.113	−0.113	−0.113	−0.113	0.486	0.486	0.486
	0.80	−0.114	−0.114	−0.114	−0.114	−0.114	−0.114	−0.114	−0.114	0.486	0.486	0.486
$\bar{\sigma}_\theta$ η_1	0	−1.2	−1.2	−1.2	−1.2	−1.2	−1.2	−1.2	−1.2	−0.6	−0.6	−0.4
	0.01	−1.129	−1.125	−1.110	−1.098	−1.090	−1.078	−1.060	−1.027	−0.427	−0.387	−0.228
	0.02	−1.058	−1.051	−1.020	−0.998	−0.982	−0.960	−0.927	−0.876	−0.276	−0.219	−0.103
	0.05	−0.853	−0.836	−0.768	−0.723	−0.692	−0.654	−0.605	−0.544	0.056	0.100	0.157
	0.10	−0.552	−0.527	−0.438	−0.337	−0.357	−0.323	−0.286	−0.246	0.354	0.364	0.378
	0.15	−0.326	−0.303	−0.228	−0.191	−0.171	−0.150	−0.128	−0.106	0.494	0.484	0.479
	0.20	−0.176	−0.159	−0.109	−0.088	−0.076	−0.065	−0.054	−0.043	0.557	0.537	0.520
	0.25	−0.088	−0.078	−0.050	−0.039	−0.034	−0.028	−0.023	−0.019	0.581	0.555	0.533
	0.30	−0.043	−0.038	−0.026	−0.021	−0.019	−0.017	−0.016	−0.014	0.586	0.556	0.531
	0.40	−0.022	−0.022	−0.023	−0.024	−0.024	−0.025	−0.025	−0.026	0.574	0.542	0.515
	0.50	−0.032	−0.033	−0.036	−0.037	−0.038	−0.039	−0.040	−0.041	0.559	0.528	0.500
	0.60	−0.044	−0.045	−0.047	−0.048	−0.048	−0.049	−0.049	−0.050	0.550	0.519	0.492
	0.70	−0.051	−0.052	−0.053	−0.054	−0.054	−0.054	−0.054	−0.055	0.545	0.514	0.488
	0.80	−0.057	−0.057	−0.057	−0.057	−0.057	−0.057	−0.057	−0.057	0.544	0.512	0.486
$\bar{\sigma}_r$ η_1	0	−1.2	−1.2	−1.2	−1.2	−1.2	−1.2	−1.2	−0.90	−0.90	−0.6	0
	0.01	−1.129	−1.120	−1.080	−1.047	−1.018	−0.972	−0.881	−0.614	−0.614	−0.250	0
	0.02	−1.058	−1.041	−0.964	−0.899	−0.847	−0.767	−0.634	−0.407	−0.407	−0.118	0
	0.05	−0.853	−0.814	−0.653	−0.537	−0.458	−0.362	−0.251	−0.143	−0.143	−0.038	0
	0.10	−0.552	−0.499	−0.315	−0.222	−0.174	−0.129	−0.092	−0.067	−0.067	−0.025	0
	0.15	−0.326	−0.280	−0.154	−0.108	−0.089	−0.074	−0.063	−0.058	−0.058	−0.026	0
	0.20	−0.176	−0.147	−0.084	−0.068	−0.063	−0.059	−0.058	−0.058	−0.058	−0.027	0
	0.25	−0.088	−0.075	−0.055	−0.054	−0.054	−0.055	−0.057	−0.058	−0.058	−0.028	0
	0.30	−0.043	−0.041	−0.045	−0.049	−0.052	−0.054	−0.057	−0.058	−0.058	−0.028	0
	0.40	−0.022	−0.028	−0.044	−0.050	−0.052	−0.055	−0.057	−0.058	−0.058	−0.027	0
	0.50	−0.032	−0.037	−0.049	−0.053	−0.054	−0.056	−0.057	−0.058	−0.058	−0.027	0
	0.60	−0.044	−0.047	−0.053	−0.055	−0.056	−0.056	−0.057	−0.057	−0.057	−0.026	0
	0.70	−0.051	−0.053	−0.055	−0.056	−0.057	−0.057	−0.057	−0.057	−0.057	−0.026	0
	0.80	−0.057	−0.057	−0.057	−0.057	−0.057	−0.057	−0.057	−0.057	−0.057	−0.026	0
$\bar{\tau}_{rz}$ η_1	0	0	0.191	0.417	0.514	0.574	0.649	0.768	$+\infty$	$+\infty$	0.662	0
	0.01	0	0.190	0.413	0.506	0.561	0.628	0.716	0.778	0.778	0.518	0
	0.02	0	0.186	0.399	0.482	0.526	0.570	0.604	0.566	0.566	0.327	0
	0.05	0	0.162	0.315	0.346	0.346	0.328	0.280	0.192	0.192	0.078	0
	0.10	0	0.095	0.131	0.107	0.084	0.055	0.023	−0.005	−0.005	−0.017	0
	0.15	0	0.026	−0.001	−0.027	−0.040	−0.050	−0.055	−0.051	−0.051	−0.034	0
	0.20	0	−0.021	−0.064	−0.077	−0.080	−0.079	−0.072	−0.057	−0.057	−0.033	0
	0.25	0	−0.043	−0.083	−0.086	−0.084	−0.077	−0.066	−0.050	−0.050	−0.028	0
	0.30	0	−0.049	−0.079	−0.077	−0.072	−0.064	−0.053	−0.039	−0.039	−0.021	0
	0.40	0	−0.036	−0.050	−0.045	−0.041	−0.035	−0.028	−0.020	−0.020	−0.011	0
	0.50	0	−0.019	−0.024	−0.021	−0.019	−0.016	−0.012	−0.009	−0.009	−0.004	0
	0.60	0	−0.008	−0.010	−0.008	−0.007	−0.006	−0.004	−0.003	−0.003	−0.001	0
	0.70	0	−0.003	−0.003	−0.003	−0.002	−0.002	−0.001	−0.001	−0.001	0.000	0
	0.80	0	0.000	0.000	0.000	0.000	0.000	0.000	0.000	0.000	0.000	0

	λ p	Core									Shell	
		0	0.3	0.6	0.7	0.75	0.80	0.85	0.90	0.95	0.95	1.00
$\dfrac{\bar{\sigma}_z}{\eta_1}$	0	0	0	0	0	0	0	0	0	−0.30	0.30	0.6
	0.01	−0.039	−0.040	−0.047	−0.050	−0.052	−0.057	−0.065	−0.089	−0.194	0.410	0.995
	0.02	−0.077	−0.080	−0.091	−0.097	−0.099	−0.103	−0.105	−0.099	−0.008	0.596	1.210
	0.05	−0.184	−0.190	−0.198	−0.188	−0.172	−0.136	−0.066	0.075	0.338	0.940	1.355
	0.10	−0.319	−0.316	−0.260	−0.187	−0.125	−0.039	0.080	0.235	0.426	1.025	1.241
	0.15	−0.382	−0.360	−0.232	−0.129	−0.059	0.025	0.124	0.234	0.354	0.954	1.076
	0.20	−0.382	−0.343	−0.184	−0.085	−0.026	0.040	0.110	0.183	0.257	0.857	0.927
	0.25	−0.341	−0.296	−0.141	−0.061	−0.016	0.030	0.077	0.124	0.168	0.768	0.808
	0.30	−0.283	−0.241	−0.110	−0.050	−0.019	0.012	0.043	0.071	0.096	0.696	0.718
	0.40	−0.176	−0.148	−0.075	−0.048	−0.034	−0.022	−0.011	−0.002	0.005	0.605	0.609
	0.50	−0.108	−0.094	−0.062	−0.052	−0.047	−0.043	−0.040	−0.038	−0.038	0.562	0.562
	0.60	−0.075	−0.070	−0.058	−0.055	−0.054	−0.053	−0.053	−0.053	−0.054	0.546	0.545
	0.70	−0.062	−0.061	−0.058	−0.057	−0.057	−0.057	−0.058	−0.058	−0.059	0.541	0.541
	0.80	−0.058	−0.058	−0.058	−0.058	−0.058	−0.058	−0.058	−0.058	−0.058	0.541	0.541
$\dfrac{\bar{\sigma}_\theta}{\eta_1}$	0	−1.2	−1.2	−1.2	−1.2	−1.2	−1.2	−1.2	−1.2	−1.2	−0.6	−0.4
	0.01	−1.126	−1.122	−1.105	−1.093	−1.084	−1.070	−1.052	−1.021	−0.953	−0.353	−0.179
	0.02	−1.052	−1.044	−1.011	−0.987	−0.970	−0.946	−0.912	−0.859	−0.765	−0.165	−0.031
	0.05	−0.837	−0.819	−0.747	−0.698	−0.665	−0.624	−0.571	−0.505	−0.424	0.176	0.239
	0.10	−0.525	−0.498	−0.403	−0.350	−0.318	−0.282	−0.243	−0.201	−0.159	0.441	0.454
	0.15	−0.291	−0.266	−0.188	−0.150	−0.129	−0.108	−0.086	−0.064	−0.042	0.558	0.549
	0.20	−0.138	−0.121	−0.070	−0.048	−0.037	−0.025	−0.014	−0.004	0.007	0.607	0.587
	0.25	−0.050	−0.040	−0.013	−0.002	0.003	0.008	0.013	0.017	0.022	0.622	0.596
	0.30	−0.007	−0.002	0.009	0.013	0.015	0.017	0.018	0.019	0.020	0.620	0.592
	0.40	0.010	0.010	0.008	0.007	0.007	0.009	0.005	0.004	0.003	0.603	0.573
	0.50	−0.002	−0.003	−0.007	−0.008	−0.009	−0.010	−0.011	−0.012	−0.013	0.587	0.557
	0.60	−0.015	−0.016	−0.019	−0.020	−0.020	−0.021	−0.021	−0.022	−0.023	0.577	0.548
	0.70	−0.023	−0.024	−0.025	−0.026	−0.026	−0.026	−0.027	−0.027	−0.027	0.573	0.543
	0.80	−0.029	−0.029	−0.029	−0.029	−0.029	−0.029	−0.029	−0.029	−0.029	0.571	0.542
$\dfrac{\bar{\sigma}_r}{\eta_1}$	0	−1.2	−1.2	−1.2	−1.2	−1.2	−1.2	−1.2	−1.2	−0.90	−0.90	0
	0.01	−1.126	−1.117	−1.075	−1.040	−1.014	−0.971	−0.904	−0.769	−0.403	−0.403	0
	0.02	−1.052	−1.034	−0.953	−0.887	−0.837	−0.762	−0.649	−0.465	−0.182	−0.182	0
	0.05	−0.837	−0.797	−0.628	−0.509	−0.428	−0.334	−0.225	−0.116	−0.043	−0.043	0
	0.10	−0.525	−0.469	−0.278	−0.185	−0.138	−0.093	−0.058	−0.035	−0.027	−0.027	0
	0.15	−0.291	−0.244	−0.117	−0.073	−0.055	−0.041	−0.033	−0.029	−0.029	−0.029	0
	0.20	−0.138	−0.110	−0.050	−0.035	−0.031	−0.029	−0.029	−0.030	−0.030	−0.030	0
	0.25	−0.050	−0.039	−0.023	−0.023	−0.025	−0.026	−0.028	−0.030	−0.031	−0.031	0
	0.30	−0.007	−0.007	−0.014	−0.020	−0.022	−0.026	−0.028	−0.030	−0.031	−0.031	0
	0.40	0.010	0.004	−0.014	−0.021	−0.024	−0.027	−0.029	−0.030	−0.030	−0.030	0
	0.50	−0.002	−0.008	−0.020	−0.025	−0.027	−0.028	−0.029	−0.030	−0.030	−0.030	0
	0.60	−0.015	−0.018	−0.025	−0.027	−0.028	−0.029	−0.029	−0.029	−0.029	−0.029	0
	0.70	−0.023	−0.025	−0.028	−0.028	−0.029	−0.029	−0.029	−0.029	−0.029	−0.029	0
	0.80	−0.029	−0.029	−0.029	−0.029	−0.029	−0.029	−0.029	−0.029	−0.029	−0.029	0
$\dfrac{\bar{\tau}_{rz}}{\eta_1}$	0	0	0.189	0.407	0.494	0.541	0.596	0.660	0.755	$+\infty$	$+\infty$	0
	0.01	0	0.188	0.402	0.486	0.530	0.578	0.628	0.648	0.633	0.633	0
	0.02	0	0.184	0.388	0.461	0.497	0.527	0.545	0.521	0.346	0.346	0
	0.05	0	0.159	0.301	0.324	0.320	0.297	0.246	0.160	0.055	0.055	0
	0.10	0	0.088	0.112	0.092	0.057	0.026	−0.005	−0.028	−0.031	−0.031	0
	0.15	0	0.018	−0.020	−0.048	−0.062	−0.071	−0.073	−0.065	−0.042	−0.042	0
	0.20	0	−0.030	−0.081	−0.094	−0.095	−0.093	−0.083	−0.065	−0.038	−0.038	0
	0.25	0	−0.051	−0.095	−0.098	−0.094	−0.086	−0.073	−0.055	−0.030	−0.030	0
	0.30	0	−0.055	−0.087	−0.084	−0.079	−0.070	−0.058	−0.042	−0.023	−0.023	0
	0.40	0	−0.039	−0.053	−0.048	−0.043	−0.037	−0.030	−0.021	−0.011	−0.011	0
	0.50	0	−0.020	−0.025	−0.022	−0.019	−0.016	−0.013	−0.009	−0.004	−0.004	0
	0.60	0	−0.008	−0.010	−0.008	−0.007	−0.006	−0.004	−0.003	−0.001	−0.001	0
	0.70	0	−0.003	−0.003	−0.002	−0.002	−0.002	−0.001	−0.001	0.000	0.000	0
	0.80	0	0.000	0.000	0.000	0.000	0.000	0.000	0.000	0.000	0.000	0

$\gamma = 0,5 \: ; \: \mu = 0,7 \: ; \: \delta = 2$

	λ / p	Core				Shell						
		0	0.3	0.6	0.7	0.7	0.75	0.80	0.85	0.90	0.95	1.00
$\bar{\sigma}_z$ η_1	0	0	0	0	0.261	−0.228	0	0	0	0	0	0.6
	0.01	−0.018	−0.017	0.002	0.245	−0.224	−0.062	−0.044	−0.040	−0.035	0.002	0.577
	0.02	−0.035	−0.033	0.009	0.231	−0.224	−0.116	−0.083	−0.068	−0.039	0.082	0.544
	0.05	−0.081	−0.074	0.036	0.203	−0.212	−0.178	−0.133	−0.073	0.026	0.193	0.426
	0.10	−0.122	−0.096	0.077	0.220	−0.186	−0.155	−0.106	−0.040	0.042	0.136	0.232
	0.15	−0.111	−0.069	0.121	0.240	−0.171	−0.137	−0.097	−0.052	−0.003	0.046	0.090
	0.20	−0.065	−0.016	0.155	0.242	−0.169	−0.141	−0.112	−0.082	−0.054	−0.028	−0.007
	0.25	−0.005	0.040	0.170	0.234	−0.173	−0.152	−0.132	−0.113	−0.097	−0.083	−0.072
	0.30	0.053	0.090	0.188	0.225	−0.178	−0.164	−0.150	−0.139	−0.129	−0.121	−0.117
	0.40	0.137	0.155	0.197	0.209	−0.186	−0.180	−0.175	−0.171	−0.168	−0.166	−0.166
	0.50	0.178	0.185	0.199	0.203	−0.190	−0.188	−0.186	−0.185	−0.184	−0.185	−0.185
	0.60	0.194	0.196	0.200	0.200	−0.191	−0.191	−0.190	−0.190	−0.190	−0.191	−0.191
	0.70	0.199	0.200	0.200	0.200	−0.192	−0.192	−0.192	−0.192	−0.192	−0.192	−0.192
	0.80	0.200	0.200	0.200	0.200	−0.192	−0.192	−0.192	−0.192	−0.192	−0.192	−0.192
$\bar{\sigma}_\theta$ η_1	0	−0.6	−0.6	−0.6	−0.554	−0.635	−0.6	−0.6	−0.6	−0.6	−0.6	−0.4
	0.01	−0.547	−0.543	−0.522	−0.473	−0.584	−0.557	−0.544	−0.531	−0.511	−0.466	−0.316
	0.02	−0.494	−0.487	−0.447	−0.399	−0.537	−0.512	−0.491	−0.468	−0.433	−0.372	−0.266
	0.05	−0.344	−0.327	−0.256	−0.208	−0.418	−0.389	−0.358	−0.323	−0.282	−0.234	−0.183
	0.10	−0.135	−0.114	−0.040	−0.002	−0.297	−0.267	−0.238	−0.210	−0.183	−0.157	−0.134
	0.15	0.006	0.023	0.072	0.094	−0.244	−0.217	−0.193	−0.172	−0.153	−0.137	−0.123
	0.20	0.085	0.095	0.120	0.130	−0.225	−0.200	−0.180	−0.163	−0.148	−0.136	−0.126
	0.25	0.120	0.125	0.134	0.137	−0.222	−0.199	−0.180	−0.165	−0.152	−0.141	−0.133
	0.30	0.130	0.131	0.132	0.131	−0.225	−0.203	−0.185	−0.170	−0.158	−0.148	−0.140
	0.40	0.119	0.118	0.114	0.112	−0.235	−0.213	−0.196	−0.182	−0.170	−0.160	−0.152
	0.50	0.103	0.102	0.099	0.097	−0.243	−0.221	−0.204	−0.189	−0.177	−0.167	−0.159
	0.60	0.093	0.092	0.090	0.090	−0.246	−0.225	−0.208	−0.193	−0.181	−0.171	−0.162
	0.70	0.088	0.087	0.087	0.087	−0.248	−0.227	−0.209	−0.195	−0.182	−0.172	−0.163
	0.80	0.085	0.085	0.085	0.085	−0.249	−0.227	−0.210	−0.195	−0.183	−0.173	−0.164
$\bar{\sigma}_r$ η_1	0	−0.6	−0.6	−0.6	−0.584	−0.584	−0.6	−0.6	−0.6	−0.6	−0.6	0
	0.01	−0.547	−0.539	−0.490	−0.480	−0.480	−0.507	−0.477	−0.437	−0.366	−0.205	0
	0.02	−0.494	−0.478	−0.394	−0.382	−0.382	−0.395	−0.359	−0.297	−0.199	−0.060	0
	0.05	−0.344	−0.310	−0.185	−0.142	−0.142	−0.132	−0.104	−0.063	−0.020	0.003	0
	0.10	−0.135	−0.096	0.008	0.042	0.042	0.037	0.033	0.027	0.020	0.009	0
	0.15	0.006	0.031	0.081	0.090	0.090	0.070	0.053	0.037	0.022	0.009	0
	0.20	0.085	0.094	0.103	0.100	0.100	0.075	0.053	0.035	0.020	0.008	0
	0.25	0.120	0.119	0.107	0.100	0.100	0.074	0.052	0.034	0.020	0.008	0
	0.30	0.130	0.124	0.104	0.097	0.097	0.072	0.051	0.033	0.019	0.008	0
	0.40	0.119	0.112	0.097	0.093	0.093	0.068	0.049	0.032	0.019	0.009	0
	0.50	0.103	0.099	0.091	0.089	0.089	0.066	0.048	0.032	0.019	0.009	0
	0.60	0.093	0.091	0.087	0.087	0.087	0.065	0.046	0.032	0.019	0.009	0
	0.70	0.088	0.087	0.086	0.085	0.085	0.064	0.046	0.031	0.019	0.009	0
	0.80	0.085	0.085	0.085	0.085	0.085	0.064	0.046	0.031	0.019	0.009	0
$\bar{\tau}_{rz}$ η_1	0	0	0.084	0.127	$-\infty$	$-\infty$	0.110	0.180	0.237	0.290	0.341	0
	0.01	0	0.083	0.119	−0.006	−0.006	0.097	0.168	0.218	0.252	0.228	0
	0.02	0	0.079	0.101	0.034	0.034	0.080	0.137	0.171	0.172	0.095	0
	0.05	0	0.057	0.042	0.022	0.022	0.021	0.024	0.017	−0.005	−0.026	0
	0.10	0	0.004	−0.044	−0.061	−0.061	−0.065	−0.070	−0.071	−0.064	−0.043	0
	0.15	0	−0.037	−0.088	−0.092	−0.092	−0.088	−0.082	−0.073	−0.057	−0.033	0
	0.20	0	−0.055	−0.094	−0.087	−0.087	−0.080	−0.071	−0.059	−0.043	−0.023	0
	0.25	0	−0.055	−0.080	−0.071	−0.071	−0.063	−0.054	−0.044	−0.031	−0.016	0
	0.30	0	−0.047	−0.062	−0.053	−0.053	−0.046	−0.039	−0.031	−0.022	−0.011	0
	0.40	0	−0.025	−0.029	−0.024	−0.024	−0.021	−0.018	−0.014	−0.010	−0.005	0
	0.50	0	−0.010	−0.012	−0.010	−0.010	−0.008	−0.007	−0.005	−0.003	−0.002	0
	0.60	0	−0.004	−0.004	−0.003	−0.003	−0.003	−0.002	−0.002	−0.001	0.000	0
	0.70	0	−0.001	−0.001	−0.001	−0.001	−0.001	0.000	0.000	0.000	0.000	0
	0.80	0	0.000	0.000	0.000	0.000	0.000	0.000	0.000	0.000	0.000	0

	λ / p	Core					Shell					
		0	0.3	0.6	0.7	0.75	0.75	0.80	0.85	0.90	0.95	1.00
$\bar{\sigma}_z$ η_1	0	0	0	0	0	0.261	−0.228	0	0	0	0	0.6
	0.01	−0.019	−0.019	−0.009	0.032	0.247	−0.223	−0.064	−0.047	−0.040	−0.003	0.572
	0.02	−0.037	−0.036	−0.015	0.066	0.233	−0.219	−0.117	−0.082	−0.049	0.072	0.533
	0.05	−0.086	−0.082	−0.012	0.113	0.222	−0.200	−0.161	−0.097	0.004	0.170	0.398
	0.10	−0.133	−0.113	0.031	0.163	0.255	−0.167	−0.124	−0.063	0.015	0.104	0.193
	0.15	−0.129	−0.093	0.082	0.197	0.266	−0.159	−0.120	−0.077	−0.031	0.014	0.054
	0.20	−0.089	−0.044	0.119	0.205	0.251	−0.166	−0.137	−0.109	−0.081	−0.057	−0.037
	0.25	−0.032	0.010	0.142	0.200	0.228	−0.177	−0.158	−0.139	−0.123	−0.109	−0.099
	0.30	0.023	0.058	0.154	0.191	0.208	−0.188	−0.175	−0.163	−0.154	−0.147	−0.142
	0.40	0.104	0.122	0.163	0.176	0.181	−0.202	−0.197	−0.193	−0.190	−0.189	−0.188
	0.50	0.144	0.151	0.165	0.168	0.169	−0.208	−0.207	−0.206	−0.205	−0.205	−0.206
	0.60	0.160	0.162	0.165	0.165	0.165	−0.210	−0.209	−0.210	−0.210	−0.210	−0.211
	0.70	0.164	0.164	0.164	0.164	0.164	−0.211	−0.211	−0.211	−0.211	−0.211	−0.212
	0.80	0.164	0.164	0.164	0.164	0.164	−0.211	−0.211	−0.211	−0.211	−0.211	−0.211
$\bar{\sigma}_\theta$ η_1	0	−0.6	−0.6	−0.6	−0.6	−0.554	−0.635	−0.6	−0.6	−0.6	−0.6	−0.4
	0.01	−0.549	−0.545	−0.527	−0.505	−0.468	−0.580	−0.551	−0.535	−0.514	−0.469	−0.318
	0.02	−0.498	−0.491	−0.457	−0.422	−0.389	−0.530	−0.501	−0.475	−0.439	−0.378	−0.272
	0.05	−0.352	−0.336	−0.269	−0.222	−0.192	−0.406	−0.374	−0.338	−0.296	−0.248	−0.197
	0.10	−0.148	−0.128	−0.055	−0.016	0.005	−0.292	−0.262	−0.232	−0.205	−0.179	−0.156
	0.15	−0.010	0.007	0.057	0.079	0.090	−0.247	−0.221	−0.198	−0.179	−0.162	−0.147
	0.20	0.069	0.079	0.105	0.115	0.119	−0.232	−0.209	−0.190	−0.174	−0.161	−0.149
	0.25	0.105	0.109	0.119	0.122	0.123	−0.230	−0.209	−0.192	−0.178	−0.166	−0.156
	0.30	0.115	0.116	0.117	0.116	0.116	−0.234	−0.214	−0.197	−0.184	−0.172	−0.163
	0.40	0.105	0.103	0.099	0.097	0.096	−0.244	−0.224	−0.208	−0.195	−0.184	−0.174
	0.50	0.089	0.088	0.085	0.083	0.083	−0.251	−0.232	−0.215	−0.202	−0.190	−0.181
	0.60	0.079	0.078	0.077	0.076	0.076	−0.255	−0.235	−0.219	−0.205	−0.193	−0.184
	0.70	0.074	0.074	0.073	0.073	0.073	−0.256	−0.236	−0.220	−0.206	−0.195	−0.185
	0.80	0.072	0.072	0.072	0.072	0.072	−0.257	−0.237	−0.220	−0.207	−0.195	−0.185
$\bar{\sigma}_r$ η_1	0	−0.6	−0.6	−0.6	−0.6	−0.584	−0.584	−0.6	−0.6	−0.6	−0.6	0
	0.01	−0.549	−0.541	−0.500	−0.451	−0.457	−0.457	−0.482	−0.439	−0.366	−0.205	0
	0.02	−0.498	−0.483	−0.406	−0.346	−0.342	−0.342	−0.352	−0.296	−0.198	−0.059	0
	0.05	−0.352	−0.319	−0.188	−0.127	−0.096	−0.096	−0.081	−0.050	−0.015	0.005	0
	0.10	−0.148	−0.109	0.002	0.039	0.053	0.053	0.046	0.037	0.025	0.011	0
	0.15	−0.010	0.017	0.070	0.079	0.081	0.081	0.050	0.041	0.024	0.010	0
	0.20	0.069	0.079	0.090	0.086	0.083	0.083	0.059	0.039	0.023	0.009	0
	0.25	0.105	0.104	0.093	0.086	0.082	0.082	0.058	0.038	0.022	0.009	0
	0.30	0.115	0.109	0.090	0.083	0.080	0.080	0.057	0.038	0.022	0.010	0
	0.40	0.105	0.098	0.083	0.078	0.076	0.076	0.055	0.037	0.022	0.010	0
	0.50	0.089	0.085	0.077	0.075	0.074	0.074	0.053	0.036	0.022	0.010	0
	0.60	0.079	0.077	0.074	0.073	0.073	0.073	0.052	0.036	0.022	0.010	0
	0.70	0.074	0.074	0.073	0.072	0.072	0.072	0.052	0.036	0.022	0.010	0
	0.80	0.072	0.072	0.072	0.072	0.072	0.072	0.052	0.036	0.022	0.010	0
$\bar{\tau}_{rz}$ η_1	0	0	0.089	0.159	0.128	$-\infty$	$-\infty$	0.137	0.211	0.274	0.332	0
	0.01	0	0.088	0.153	0.110	0.023	0.023	0.121	0.191	0.236	0.220	0
	0.02	0	0.085	0.138	0.088	0.054	0.054	0.095	0.144	0.155	0.087	0
	0.05	0	0.064	0.069	0.032	0.019	0.019	0.010	0.001	−0.017	−0.033	0
	0.10	0	0.011	−0.035	−0.061	−0.069	−0.069	−0.072	−0.073	−0.066	−0.043	0
	0.15	0	−0.031	−0.085	−0.093	−0.089	−0.089	−0.082	−0.071	−0.055	−0.031	0
	0.20	0	−0.051	−0.092	−0.088	−0.079	−0.079	−0.069	−0.057	−0.041	−0.022	0
	0.25	0	−0.053	−0.079	−0.070	−0.061	−0.061	−0.052	−0.042	−0.030	−0.016	0
	0.30	0	−0.045	−0.060	−0.052	−0.044	−0.044	−0.037	−0.030	−0.021	−0.011	0
	0.40	0	−0.024	−0.029	−0.023	−0.019	−0.019	−0.016	−0.013	−0.009	−0.004	0
	0.50	0	−0.010	−0.011	−0.009	−0.007	−0.007	−0.006	−0.005	−0.003	−0.002	0
	0.60	0	−0.003	−0.003	−0.002	−0.002	−0.002	−0.002	−0.001	−0.001	0.000	0
	0.70	0	−0.001	−0.001	0.000	0.000	0.000	0.000	0.000	0.000	0.000	0
	0.80	0	0.000	0.000	0.000	0.000	0.000	0.000	0.000	0.000	0.000	0

$\gamma = 0,5$; $\mu = 0,8$; $\delta = 2$

	λ	Core						Shell				
	p	0	0.3	0.6	0.7	0.75	0.80	0.80	0.85	0.90	0.95	1.00
$\bar{\sigma}_z$ η_1	0	0	0	0	0	0	0.261	-0.228	0	0	0	0.6
	0.01	-0.019	-0.020	-0.015	-0.001	0.031	0.248	-0.220	-0.066	-0.048	-0.010	0.564
	0.02	-0.039	-0.039	-0.028	0.004	0.065	0.239	-0.212	-0.116	-0.065	0.058	0.517
	0.05	-0.090	-0.088	-0.044	0.039	0.125	0.253	-0.179	-0.126	-0.025	0.138	0.357
	0.10	-0.143	-0.128	-0.010	0.107	0.193	0.299	-0.144	-0.089	-0.017	0.065	0.146
	0.15	-0.145	-0.114	0.043	0.150	0.217	0.290	-0.148	-0.106	-0.063	-0.020	0.017
	0.20	-0.111	-0.070	0.082	0.164	0.209	0.255	-0.166	-0.138	-0.111	-0.088	-0.068
	0.25	-0.059	-0.019	0.106	0.162	0.191	0.218	-0.185	-0.166	-0.150	-0.137	-0.127
	0.30	-0.007	0.027	0.119	0.155	0.173	0.188	-0.200	-0.188	-0.179	-0.172	-0.167
	0.40	0.071	0.088	0.129	0.142	0.147	0.151	-0.218	-0.215	-0.212	-0.211	-0.210
	0.50	0.110	0.117	0.130	0.134	0.135	0.135	-0.227	-0.226	-0.225	-0.225	-0.226
	0.60	0.125	0.127	0.130	0.130	0.130	0.130	-0.229	-0.229	-0.229	-0.230	-0.230
	0.70	0.129	0.129	0.129	0.129	0.129	0.128	-0.230	-0.230	-0.230	-0.230	-0.231
	0.80	0.129	0.129	0.129	0.129	0.129	0.129	-0.230	-0.230	-0.230	-0.230	-0.230
$\bar{\sigma}_\theta$ η_1	0	-0.6	-0.6	-0.6	-0.6	-0.6	-0.554	-0.635	-0.6	-0.6	-0.6	-0.4
	0.01	-0.551	-0.547	-0.532	-0.517	-0.501	-0.461	-0.574	-0.542	-0.518	-0.472	-0.322
	0.02	-0.502	-0.495	-0.465	-0.437	-0.413	-0.376	-0.518	-0.485	-0.447	-0.386	-0.280
	0.05	-0.361	-0.346	-0.284	-0.239	-0.209	-0.172	-0.391	-0.355	-0.313	-0.265	-0.215
	0.10	-0.164	-0.143	-0.073	-0.035	-0.013	0.011	-0.289	-0.259	-0.230	-0.204	-0.181
	0.15	-0.028	-0.011	0.038	0.060	0.072	0.083	-0.252	-0.227	-0.206	-0.188	-0.173
	0.20	0.050	0.060	0.086	0.096	0.101	0.106	-0.241	-0.219	-0.202	-0.187	-0.174
	0.25	0.087	0.091	0.101	0.104	0.106	0.107	-0.240	-0.221	-0.205	-0.191	-0.180
	0.30	0.098	0.099	0.100	0.100	0.100	0.099	-0.244	-0.225	-0.210	-0.197	-0.186
	0.40	0.089	0.088	0.084	0.082	0.082	0.081	-0.254	-0.236	-0.220	-0.208	-0.197
	0.50	0.074	0.073	0.070	0.069	0.068	0.068	-0.260	-0.242	-0.227	-0.214	-0.203
	0.60	0.065	0.064	0.063	0.062	0.062	0.061	-0.263	-0.245	-0.230	-0.217	-0.206
	0.70	0.060	0.060	0.059	0.059	0.059	0.059	-0.265	-0.246	-0.231	-0.218	-0.207
	0.80	0.058	0.058	0.058	0.058	0.058	0.058	-0.265	-0.247	-0.231	-0.218	-0.207
$\bar{\sigma}_r$ η_1	0	-0.6	-0.6	-0.6	-0.6	-0.6	-0.584	-0.584	-0.6	-0.6	-0.6	0
	0.01	-0.551	-0.543	-0.506	-0.469	-0.434	-0.437	-0.437	-0.442	-0.367	-0.205	0
	0.02	-0.502	-0.487	-0.417	-0.355	-0.315	-0.305	-0.305	-0.287	-0.197	-0.059	0
	0.05	-0.361	-0.329	-0.199	-0.123	-0.085	-0.050	-0.050	-0.028	-0.004	0.008	0
	0.10	-0.164	-0.124	-0.009	0.032	0.048	0.058	0.058	0.045	0.030	0.013	0
	0.15	-0.028	-0.001	0.056	0.067	0.068	0.067	0.067	0.046	0.027	0.011	0
	0.20	0.050	0.062	0.075	0.072	0.069	0.066	0.066	0.044	0.025	0.011	0
	0.25	0.087	0.087	0.077	0.071	0.067	0.065	0.065	0.043	0.025	0.011	0
	0.30	0.098	0.093	0.075	0.068	0.065	0.063	0.063	0.042	0.025	0.011	0
	0.40	0.089	0.083	0.068	0.064	0.062	0.061	0.061	0.041	0.024	0.011	0
	0.50	0.074	0.071	0.063	0.061	0.060	0.059	0.059	0.040	0.024	0.011	0
	0.60	0.065	0.063	0.060	0.059	0.059	0.058	0.058	0.040	0.024	0.011	0
	0.70	0.060	0.060	0.059	0.058	0.058	0.058	0.058	0.040	0.024	0.011	0
	0.80	0.058	0.058	0.058	0.058	0.058	0.058	0.058	0.040	0.024	0.011	0
$\bar{\tau}_{rz}$ η_1	0	0	0.094	0.181	0.186	0.160	$-\infty$	$-\infty$	0.167	0.249	0.320	0
	0.01	0	0.093	0.176	0.176	0.140	0.046	0.046	0.144	0.210	0.207	0
	0.02	0	0.090	0.163	0.151	0.112	0.072	0.072	0.103	0.129	0.074	0
	0.05	0	0.069	0.093	0.059	0.031	0.007	0.007	-0.011	-0.030	-0.041	0
	0.10	0	0.018	-0.022	-0.054	-0.067	-0.074	-0.074	-0.072	-0.064	-0.042	0
	0.15	0	-0.025	-0.078	-0.090	-0.089	-0.080	-0.080	-0.068	-0.052	-0.029	0
	0.20	0	-0.047	-0.088	-0.086	-0.078	-0.065	-0.065	-0.054	-0.039	-0.021	0
	0.25	0	-0.050	-0.076	-0.069	-0.061	-0.049	-0.049	-0.039	-0.028	-0.015	0
	0.30	0	-0.043	-0.059	-0.051	-0.044	-0.034	-0.034	-0.027	-0.019	-0.010	0
	0.40	0	-0.024	-0.028	-0.022	-0.019	-0.014	-0.014	-0.012	-0.008	-0.004	0
	0.50	0	-0.010	-0.011	-0.008	-0.007	-0.005	-0.005	-0.004	-0.003	-0.001	0
	0.60	0	-0.003	-0.003	-0.002	-0.002	-0.001	-0.001	-0.001	-0.001	0.000	0
	0.70	0	-0.001	-0.001	0.000	0.000	0.000	0.000	0.000	0.000	0.000	0
	0.80	0	0.000	0.000	0.000	0.000	0.000	0.000	0.000	0.000	0.000	0

	λ	Core							Shell			
	p	0	0.3	0.6	0.7	0.75	0.80	0.85	0.85	0.90	0.95	1.00
$\begin{array}{c}\bar{\sigma}_z\\ \eta_1^z\end{array}$	0	0	0	0	0	0	0	0.261	−0.228	0	0	0.6
	0.01	−0.020	−0.021	−0.019	−0.013	−0.003	0.030	0.250	−0.212	−0.068	−0.023	0.551
	0.02	−0.040	−0.041	−0.037	−0.022	0.001	0.066	0.251	−0.197	−0.101	0.034	0.488
	0.05	−0.093	−0.093	−0.067	−0.013	0.046	0.148	0.307	−0.147	−0.062	0.094	0.296
	0.10	−0.150	−0.139	−0.045	0.054	0.130	0.228	0.348	−0.121	−0.054	0.021	0.093
	0.15	−0.159	−0.133	0.005	0.103	0.165	0.234	0.307	−0.141	−0.099	−0.059	−0.023
	0.20	−0.131	−0.095	0.044	0.121	0.164	0.208	0.252	−0.169	−0.143	−0.121	−0.102
	0.25	−0.085	−0.048	0.069	0.123	0.151	0.178	0.203	−0.194	−0.178	−0.166	−0.156
	0.30	−0.036	−0.005	0.083	0.119	0.136	0.152	0.165	−0.213	−0.204	−0.197	−0.192
	0.40	0.037	0.054	0.094	0.107	0.113	0.117	0.120	−0.235	−0.233	−0.231	−0.231
	0.50	0.076	0.082	0.096	0.100	0.101	0.102	0.102	−0.245	−0.244	−0.244	−0.245
	0.60	0.091	0.093	0.096	0.097	0.097	0.096	0.096	−0.248	−0.248	−0.248	−0.249
	0.70	0.095	0.096	0.096	0.095	0.095	0.095	0.095	−0.248	−0.249	−0.249	−0.249
	0.80	0.095	0.095	0.095	0.095	0.095	0.095	0.095	−0.248	−0.248	−0.248	−0.248
$\begin{array}{c}\bar{\sigma}_\theta\\ \eta_1^\theta\end{array}$	0	−0.6	−0.6	−0.6	−0.6	−0.6	−0.6	−0.554	−0.635	−0.6	−0.6	−0.4
	0.01	−0.553	−0.550	−0.535	−0.523	−0.513	−0.495	−0.449	−0.564	−0.526	−0.478	−0.327
	0.02	−0.506	−0.500	−0.472	−0.449	−0.431	−0.402	−0.356	−0.501	−0.459	−0.396	−0.291
	0.05	−0.371	−0.357	−0.299	−0.258	−0.229	−0.193	−0.150	−0.375	−0.334	−0.288	−0.240
	0.10	−0.181	−0.161	−0.094	−0.057	−0.035	−0.012	0.012	−0.289	−0.260	−0.233	−0.210
	0.15	−0.048	−0.032	0.016	0.038	0.049	0.061	0.071	−0.259	−0.237	−0.217	−0.200
	0.20	0.029	0.039	0.066	0.076	0.081	0.085	0.089	−0.251	−0.231	−0.215	−0.201
	0.25	0.067	0.071	0.082	0.085	0.087	0.088	0.089	−0.251	−0.233	−0.218	−0.205
	0.30	0.079	0.080	0.083	0.083	0.082	0.082	0.082	−0.255	−0.237	−0.223	−0.211
	0.40	0.073	0.072	0.069	0.067	0.066	0.065	0.064	−0.263	−0.247	−0.232	−0.221
	0.50	0.059	0.058	0.056	0.055	0.054	0.053	0.053	−0.269	−0.252	−0.238	−0.226
	0.60	0.050	0.050	0.048	0.048	0.048	0.047	0.047	−0.272	−0.255	−0.241	−0.228
	0.70	0.046	0.046	0.045	0.045	0.045	0.045	0.045	−0.273	−0.256	−0.242	−0.229
	0.80	0.044	0.044	0.044	0.044	0.044	0.044	0.044	−0.274	−0.256	−0.242	−0.229
$\begin{array}{c}\bar{\sigma}_r\\ \eta_1^r\end{array}$	0	−0.6	−0.6	−0.6	−0.6	−0.6	−0.6	−0.584	−0.584	−0.6	−0.6	0
	0.01	−0.553	−0.546	−0.512	−0.480	−0.453	−0.408	−0.393	−0.393	−0.369	−0.205	0
	0.02	−0.506	−0.492	−0.427	−0.370	−0.325	−0.270	−0.235	−0.235	−0.185	−0.058	0
	0.05	−0.371	−0.341	−0.215	−0.132	−0.085	−0.040	−0.002	−0.002	0.013	0.014	0
	0.10	−0.181	−0.143	−0.025	0.019	0.037	0.048	0.051	0.051	0.033	0.015	0
	0.15	−0.048	−0.021	0.039	0.051	0.053	0.053	0.050	0.050	0.029	0.013	0
	0.20	0.029	0.042	0.058	0.056	0.054	0.051	0.048	0.048	0.028	0.012	0
	0.25	0.067	0.068	0.061	0.055	0.052	0.049	0.047	0.047	0.028	0.012	0
	0.30	0.079	0.075	0.059	0.053	0.050	0.048	0.046	0.046	0.027	0.012	0
	0.40	0.073	0.067	0.053	0.049	0.047	0.046	0.045	0.045	0.027	0.012	0
	0.50	0.059	0.056	0.048	0.046	0.045	0.045	0.044	0.044	0.027	0.012	0
	0.60	0.050	0.049	0.046	0.045	0.045	0.044	0.044	0.044	0.027	0.012	0
	0.70	0.046	0.046	0.045	0.044	0.044	0.044	0.044	0.044	0.027	0.012	0
	0.80	0.044	0.044	0.044	0.044	0.044	0.044	0.044	0.044	0.027	0.012	0
$\begin{array}{c}\bar{\tau}_{rz}\\ \eta_1^{rz}\end{array}$	0	0	0.097	0.197	0.220	0.221	0.197	$-\infty$	$-\infty$	0.205	0.299	0
	0.01	0	0.096	0.192	0.212	0.208	0.173	0.077	0.077	0.164	0.186	0
	0.02	0	0.093	0.180	0.190	0.177	0.134	0.081	0.081	0.090	0.054	0
	0.05	0	0.074	0.113	0.088	0.060	0.023	−0.014	−0.014	−0.038	−0.047	0
	0.10	0	0.024	−0.007	−0.041	−0.058	−0.070	−0.070	−0.070	−0.060	−0.038	0
	0.15	0	−0.019	−0.068	−0.083	−0.084	−0.078	−0.063	−0.063	−0.048	−0.027	0
	0.20	0	−0.041	−0.081	−0.082	−0.076	−0.065	−0.049	−0.049	−0.036	−0.019	0
	0.25	0	−0.046	−0.073	−0.067	−0.060	−0.049	−0.035	−0.035	−0.025	−0.013	0
	0.30	0	−0.041	−0.057	−0.050	−0.044	−0.035	−0.024	−0.024	−0.017	−0.009	0
	0.40	0	−0.023	−0.028	−0.023	−0.019	−0.015	−0.010	−0.010	−0.007	−0.004	0
	0.50	0	−0.010	−0.011	−0.008	−0.007	−0.005	−0.003	−0.003	−0.002	−0.001	0
	0.60	0	−0.003	−0.003	−0.002	−0.002	−0.001	−0.001	−0.001	−0.001	0.000	0
	0.70	0	−0.001	−0.001	0.000	0.000	0.000	0.000	0.000	0.000	0.000	0
	0.80	0	0.000	0.000	0.000	0.000	0.000	0.000	0.000	0.000	0.000	0

$\gamma = 0{,}5; \mu = 0{,}9; \delta = 2$

	λ	Core								Shell		
	p	0	0.3	0.6	0.7	0.75	0.80	0.85	0.90	0.90	0.95	1.00
$\bar\sigma_z/\eta_1$	0	0	0	0	0	0	0	0	0.261	-0.228	0	0.6
	0.01	-0.020	-0.021	-0.022	-0.020	-0.016	-0.007	0.028	0.264	-0.202	-0.050	0.524
	0.02	-0.040	-0.041	-0.043	-0.037	-0.027	-0.003	0.069	0.288	-0.169	-0.015	0.431
	0.05	-0.095	-0.096	-0.083	-0.049	-0.010	0.061	0.187	0.393	-0.100	0.034	0.208
	0.10	-0.156	-0.148	-0.075	0.006	0.071	0.156	0.263	0.388	-0.102	-0.033	0.034
	0.15	-0.171	-0.149	-0.031	0.055	0.111	0.174	0.243	0.314	-0.140	-0.101	-0.067
	0.20	-0.149	-0.118	0.006	0.077	0.117	0.159	0.201	0.241	-0.176	-0.154	-0.136
	0.25	-0.109	-0.076	0.032	0.083	0.109	0.136	0.161	0.183	-0.205	-0.193	-0.183
	0.30	-0.065	-0.036	0.047	0.081	0.098	0.114	0.128	0.140	-0.227	-0.220	-0.216
	0.40	0.004	0.020	0.060	0.073	0.079	0.084	0.088	0.090	-0.252	-0.251	-0.250
	0.50	0.041	0.048	0.063	0.067	0.068	0.069	0.070	0.070	-0.262	-0.262	-0.263
	0.60	0.057	0.059	0.063	0.064	0.064	0.064	0.064	0.063	-0.266	-0.266	-0.266
	0.70	0.062	0.062	0.063	0.063	0.063	0.062	0.062	0.062	-0.266	-0.266	-0.267
	0.80	0.062	0.062	0.062	0.062	0.062	0.062	0.062	0.062	-0.266	-0.266	-0.266
$\bar\sigma_\theta/\eta_1$	0	-0.6	-0.6	-0.6	-0.6	-0.6	-0.6	-0.6	-0.554	-0.635	-0.6	-0.4
	0.01	-0.555	-0.552	-0.539	-0.529	-0.521	-0.508	-0.485	-0.429	-0.546	-0.488	-0.336
	0.02	-0.511	-0.505	-0.480	-0.460	-0.445	-0.422	-0.385	-0.324	-0.475	-0.412	-0.310
	0.05	-0.382	-0.369	-0.316	-0.279	-0.253	-0.220	-0.178	-0.128	-0.361	-0.317	-0.273
	0.10	-0.200	-0.182	-0.119	-0.083	-0.063	-0.041	-0.017	0.006	-0.293	-0.266	-0.242
	0.15	-0.071	-0.056	-0.009	0.012	0.023	0.034	0.045	0.056	-0.269	-0.248	-0.230
	0.20	0.006	0.016	0.042	0.052	0.057	0.062	0.066	0.070	-0.262	-0.244	-0.228
	0.25	0.045	0.049	0.061	0.065	0.066	0.068	0.069	0.070	-0.262	-0.245	-0.231
	0.30	0.059	0.061	0.064	0.064	0.064	0.064	0.064	0.064	-0.265	-0.249	-0.235
	0.40	0.056	0.056	0.053	0.051	0.051	0.050	0.049	0.048	-0.273	-0.257	-0.244
	0.50	0.044	0.044	0.041	0.040	0.039	0.039	0.038	0.038	-0.278	-0.263	-0.249
	0.60	0.036	0.036	0.034	0.034	0.033	0.033	0.033	0.032	-0.281	-0.265	-0.252
	0.70	0.032	0.032	0.031	0.031	0.031	0.031	0.030	0.030	-0.282	-0.266	-0.252
	0.80	0.030	0.030	0.030	0.030	0.030	0.030	0.030	0.030	-0.282	-0.266	-0.253
$\bar\sigma_r/\eta_1$	0	-0.6	-0.6	-0.6	-0.6	-0.6	-0.6	-0.6	-0.584	-0.584	-0.6	0
	0.01	-0.555	-0.549	-0.517	-0.489	-0.466	-0.429	-0.367	-0.318	-0.318	-0.204	0
	0.02	-0.511	-0.498	-0.437	-0.385	-0.344	-0.285	-0.205	-0.131	-0.131	-0.046	0
	0.05	-0.382	-0.354	-0.235	-0.153	-0.102	-0.048	0.000	0.029	0.029	0.020	0
	0.10	-0.200	-0.164	-0.049	-0.001	0.018	0.032	0.037	0.036	0.036	0.015	0
	0.15	-0.071	-0.044	0.018	0.032	0.036	0.036	0.035	0.032	0.032	0.014	0
	0.20	0.006	0.019	0.039	0.039	0.037	0.035	0.032	0.031	0.031	0.013	0
	0.25	0.045	0.047	0.044	0.039	0.036	0.033	0.031	0.030	0.030	0.013	0
	0.30	0.059	0.056	0.043	0.037	0.034	0.032	0.030	0.030	0.030	0.013	0
	0.40	0.056	0.051	0.038	0.034	0.032	0.031	0.030	0.030	0.030	0.013	0
	0.50	0.044	0.041	0.034	0.032	0.031	0.030	0.030	0.030	0.030	0.014	0
	0.60	0.036	0.035	0.031	0.030	0.030	0.030	0.030	0.030	0.030	0.014	0
	0.70	0.032	0.031	0.030	0.030	0.030	0.030	0.030	0.030	0.030	0.014	0
	0.80	0.030	0.030	0.030	0.030	0.030	0.030	0.030	0.030	0.030	0.014	0
$\bar\tau_{rz}/\eta_1$	0	0	0.099	0.206	0.242	0.255	0.261	0.242	$-\infty$	$-\infty$	0.258	0
	0.01	0	0.098	0.203	0.235	0.245	0.244	0.209	0.100	0.100	0.144	0
	0.02	0	0.095	0.192	0.215	0.216	0.202	0.150	0.066	0.066	0.021	0
	0.05	0	0.077	0.128	0.115	0.092	0.055	0.005	-0.038	-0.038	-0.045	0
	0.10	0	0.030	0.009	-0.023	-0.041	-0.056	-0.062	-0.052	-0.052	-0.033	0
	0.15	0	-0.012	-0.055	-0.071	-0.075	-0.073	-0.062	-0.041	-0.041	-0.023	0
	0.20	0	-0.035	-0.073	-0.076	-0.072	-0.064	-0.050	-0.031	-0.031	-0.017	0
	0.25	0	-0.041	-0.068	-0.064	-0.059	-0.050	-0.037	-0.022	-0.022	-0.012	0
	0.30	0	-0.038	-0.054	-0.049	-0.044	-0.036	-0.026	-0.015	-0.015	-0.008	0
	0.40	0	-0.022	-0.028	-0.023	-0.020	-0.016	-0.011	-0.006	-0.006	-0.003	0
	0.50	0	-0.010	-0.011	-0.009	-0.008	-0.006	-0.004	-0.002	-0.002	-0.001	0
	0.60	0	-0.003	-0.004	-0.003	-0.002	-0.002	-0.001	0.000	0.000	0.000	0
	0.70	0	-0.001	-0.001	0.000	0.000	0.000	0.000	0.000	0.000	0.000	0
	0.80	0	0.000	0.000	0.000	0.000	0.000	0.000	0.000	0.000	0.000	0

	λ	Core									Shell	
	p	0	0.3	0.6	0.7	0.75	0.80	0.85	0.90	0.95	0.95	1.00
$\dfrac{\bar{\sigma}_z}{\eta_1}$	0	0	0	0	0	0	0	0	0	0.261	-0.228	0.6
	0.01	-0.020	-0.021	-0.023	-0.024	-0.022	-0.021	-0.013	0.023	0.313	-0.159	0.438
	0.02	-0.040	-0.041	-0.046	-0.044	-0.040	-0.033	-0.007	0.082	0.413	-0.095	0.285
	0.05	-0.094	-0.097	-0.093	-0.073	-0.050	-0.022	0.085	0.243	0.490	-0.053	0.093
	0.10	-0.158	-0.153	-0.099	-0.036	0.016	0.085	0.175	0.284	0.404	-0.096	-0.034
	0.15	-0.180	-0.162	-0.065	0.008	0.056	0.112	0.174	0.240	0.306	-0.146	-0.113
	0.20	-0.166	-0.139	-0.031	0.032	0.068	0.107	0.147	0.187	0.224	-0.186	-0.168
	0.25	-0.132	-0.103	-0.006	0.041	0.067	0.092	0.117	0.140	0.161	-0.218	-0.208
	0.30	-0.093	-0.067	0.010	0.043	0.059	0.076	0.091	0.104	0.115	-0.241	-0.236
	0.40	-0.029	-0.013	0.025	0.039	0.046	0.051	0.056	0.059	0.061	-0.269	-0.267
	0.50	0.008	0.015	0.030	0.035	0.037	0.038	0.039	0.039	0.039	-0.279	-0.279
	0.60	0.024	0.026	0.031	0.032	0.033	0.033	0.032	0.032	0.031	-0.283	-0.283
	0.70	0.030	0.030	0.031	0.031	0.031	0.031	0.031	0.031	0.031	-0.283	-0.283
	0.80	0.031	0.031	0.031	0.031	0.031	0.031	0.031	0.031	0.031	-0.283	-0.283
$\dfrac{\bar{\sigma}_\theta}{\eta_1}$	0.	-0.6	-0.6	-0.6	-0.6	-0.6	-0.6	-0.6	-0.6	-0.554	-0.635	-0.4
	0.01	-0.558	-0.555	-0.544	-0.535	-0.529	-0.518	-0.502	-0.470	-0.383	-0.506	-0.358
	0.02	-0.516	-0.511	-0.489	-0.472	-0.459	-0.440	-0.412	-0.363	-0.273	-0.437	-0.350
	0.05	-0.395	-0.383	-0.336	-0.303	-0.280	-0.252	-0.215	-0.169	-0.117	-0.356	-0.315
	0.10	-0.222	-0.205	-0.148	-0.115	-0.096	-0.075	-0.053	-0.030	-0.008	-0.303	-0.277
	0.15	-0.097	-0.083	-0.039	-0.018	-0.007	0.004	0.015	0.025	0.035	-0.281	-0.261
	0.20	-0.020	-0.010	0.016	0.026	0.031	0.036	0.041	0.046	0.050	-0.274	-0.257
	0.25	0.021	0.026	0.038	0.042	0.044	0.046	0.048	0.049	0.051	-0.273	-0.258
	0.30	0.038	0.040	0.044	0.045	0.045	0.046	0.046	0.046	0.046	-0.276	-0.261
	0.40	0.039	0.039	0.037	0.036	0.035	0.034	0.034	0.033	0.032	-0.282	-0.268
	0.50	0.030	0.029	0.026	0.025	0.025	0.024	0.024	0.023	0.023	-0.287	-0.272
	0.60	0.025	0.021	0.020	0.019	0.019	0.019	0.018	0.018	0.018	-0.290	-0.275
	0.70	0.019	0.017	0.017	0.016	0.016	0.016	0.016	0.016	0.016	-0.291	-0.276
	0.80	0.015	0.015	0.015	0.015	0.015	0.015	0.015	0.015	0.015	-0.291	-0.276
$\dfrac{\bar{\sigma}_r}{\eta_1}$	0	-0.6	-0.6	-0.6	-0.6	-0.6	-0.6	-0.6	-0.6	-0.584	-0.584	0
	0.01	-0.558	-0.552	-0.524	-0.500	-0.481	-0.448	-0.395	-0.295	-0.143	-0.143	0
	0.02	-0.516	-0.505	-0.451	-0.405	-0.370	-0.314	-0.233	-0.115	-0.004	-0.004	0
	0.05	-0.395	-0.370	-0.261	-0.185	-0.134	-0.080	-0.024	0.016	0.021	0.021	0
	0.10	-0.222	-0.188	-0.079	-0.030	-0.010	0.008	0.018	0.020	0.017	0.017	0
	0.15	-0.097	-0.071	-0.008	0.010	0.015	0.018	0.018	0.016	0.015	0.015	0
	0.20	-0.020	-0.006	0.017	0.020	0.020	0.018	0.016	0.015	0.015	0.015	0
	0.25	0.021	0.025	0.025	0.022	0.019	0.017	0.016	0.014	0.015	0.015	0
	0.30	0.038	0.036	0.026	0.022	0.019	0.017	0.015	0.014	0.015	0.015	0
	0.40	0.039	0.035	0.023	0.019	0.017	0.016	0.015	0.015	0.015	0.015	0
	0.50	0.030	0.026	0.019	0.017	0.016	0.015	0.015	0.015	0.015	0.015	0
	0.60	0.025	0.020	0.017	0.016	0.015	0.015	0.015	0.015	0.015	0.015	0
	0.70	0.019	0.017	0.016	0.015	0.015	0.015	0.015	0.015	0.015	0.015	0
	0.80	0.015	0.015	0.015	0.015	0.015	0.015	0.015	0.015	0.015	0.015	0
$\dfrac{\bar{\tau}_{rz}}{\eta_1}$	0	0	0.098	0.209	0.251	0.270	0.292	0.306	0.300	$-\infty$	$-\infty$	0
	0.01	0	0.097	0.206	0.245	0.262	0.278	0.281	0.244	0.079	0.079	0
	0.02	0	0.095	0.196	0.228	0.239	0.240	0.220	0.147	0.003	0.003	0
	0.05	0	0.079	0.139	0.136	0.121	0.091	0.045	-0.009	-0.033	-0.033	0
	0.10	0	0.035	0.026	0.000	-0.017	-0.035	-0.047	-0.046	-0.024	-0.024	0
	0.15	0	-0.004	-0.040	-0.057	-0.063	-0.063	-0.057	-0.043	-0.018	-0.018	0
	0.20	0	-0.028	-0.063	-0.067	-0.065	-0.060	-0.050	-0.035	-0.014	-0.014	0
	0.25	0	-0.036	-0.062	-0.060	-0.056	-0.049	-0.039	-0.026	-0.010	-0.010	0
	0.30	0	-0.034	-0.052	-0.048	-0.044	-0.037	-0.029	-0.019	-0.007	-0.007	0
	0.40	0	-0.021	-0.028	-0.024	-0.022	-0.018	-0.013	-0.008	-0.003	-0.003	0
	0.50	0	-0.010	-0.012	-0.010	-0.009	-0.007	-0.005	-0.003	-0.001	-0.001	0
	0.60	0	-0.004	-0.004	-0.003	-0.003	-0.002	-0.002	-0.001	0.000	0.000	0
	0.70	0	-0.001	-0.001	-0.001	0.000	0.000	0.000	0.000	0.000	0.000	0
	0.80	0	0.000	0.000	0.000	0.000	0.000	0.000	0.000	0.000	0.000	0

$\gamma = 1,0 \; ; \; \mu = 0,7 \; ; \; \delta = 2$

	λ p	Core				Shell						
		0	0.3	0.6	0.7	0.7	0.75	0.80	0.85	0.90	0.95	1.00
$\bar{\sigma}_z / \eta_1$	0	0	0	0	0.168	0.003	0	0	0	0	0	0.6
	0.01	−0.042	−0.045	−0.058	0.137	−0.003	0.014	−0.003	−0.010	−0.008	0.030	0.610
	0.02	−0.084	−0.089	−0.095	0.109	−0.005	0.005	−0.006	−0.009	0.014	0.137	0.608
	0.05	−0.198	−0.202	−0.127	0.060	−0.004	−0.009	0.007	0.051	0.145	0.319	0.571
	0.10	−0.324	−0.304	−0.116	0.055	0.017	0.035	0.075	0.139	0.227	0.334	0.453
	0.15	−0.359	−0.312	−0.082	0.066	0.030	0.061	0.102	0.152	0.210	0.271	0.333
	0.20	−0.328	−0.268	−0.050	0.062	0.031	0.062	0.095	0.129	0.166	0.202	0.235
	0.25	−0.266	−0.208	−0.029	0.050	0.026	0.051	0.075	0.099	0.122	0.143	0.161
	0.30	−0.198	−0.149	−0.016	0.036	0.019	0.037	0.055	0.070	0.085	0.097	0.106
	0.40	−0.091	−0.065	−0.004	0.015	0.008	0.017	0.024	0.031	0.035	0.039	0.040
	0.50	−0.034	−0.023	−0.001	0.005	0.003	0.006	0.009	0.011	0.012	0.013	0.012
	0.60	−0.010	−0.006	0.000	0.001	0.001	0.002	0.003	0.003	0.003	0.003	0.002
	0.70	−0.002	−0.001	0.000	0.000	0.000	0.000	0.000	0.000	0.000	0.000	0.000
	0.80	0.000	0.000	0.000	0.000	0.000	0.000	0.000	0.000	0.000	0.000	0.000
$\bar{\sigma}_\theta / \eta_1$	0	−1.2	−1.2	−1.2	−1.108	−0.635	−0.6	−0.6	−0.6	−0.6	−0.6	−0.4
	0.01	−1.113	−1.107	−1.067	−0.960	−0.551	−0.534	−0.526	−0.515	−0.495	−0.450	−0.300
	0.02	−1.027	−1.014	−0.941	−0.838	−0.479	−0.468	−0.454	−0.435	−0.402	−0.342	−0.236
	0.05	−0.781	−0.753	−0.627	−0.542	−0.305	−0.292	−0.273	−0.245	−0.208	−0.162	−0.112
	0.10	−0.437	−0.402	−0.282	−0.222	−0.122	−0.110	−0.095	−0.076	−0.056	−0.034	−0.013
	0.15	−0.201	−0.173	−0.092	−0.057	−0.031	−0.023	−0.015	−0.005	0.005	0.014	0.024
	0.20	−0.061	−0.044	0.000	0.018	0.009	0.013	0.018	0.022	0.027	0.031	0.034
	0.25	0.010	0.018	0.038	0.045	0.023	0.026	0.028	0.029	0.031	0.032	0.033
	0.30	0.039	0.041	0.047	0.049	0.026	0.027	0.027	0.028	0.028	0.028	0.028
	0.40	0.040	0.039	0.035	0.033	0.018	0.018	0.017	0.017	0.017	0.016	0.015
	0.50	0.023	0.022	0.018	0.017	0.009	0.009	0.008	0.008	0.008	0.007	0.007
	0.60	0.011	0.010	0.008	0.007	0.004	0.004	0.003	0.003	0.003	0.003	0.002
	0.70	0.004	0.004	0.003	0.002	0.001	0.001	0.001	0.001	0.001	0.001	0.001
	0.80	0.000	0.000	0.000	0.000	0.000	0.000	0.000	0.000	0.000	0.000	0.000
$\bar{\sigma}_r / \eta_1$	0	−1.2	−1.2	−1.2	−0.815	−0.815	−0.6	−0.6	−0.6	−0.6	−0.6	0
	0.01	−1.113	−1.098	−0.981	−0.710	−0.710	−0.505	−0.507	−0.451	−0.372	−0.207	0
	0.02	−1.027	−0.998	−0.798	−0.602	−0.602	−0.261	−0.413	−0.324	−0.211	−0.063	0
	0.05	−0.781	−0.719	−0.455	−0.336	−0.336	−0.069	−0.186	−0.109	−0.043	−0.004	0
	0.10	−0.437	−0.367	−0.173	−0.106	−0.106	−0.015	−0.040	−0.018	−0.004	0.000	0
	0.15	−0.201	−0.154	−0.053	−0.027	−0.027	0.001	−0.007	−0.002	0.000	0.000	0
	0.20	−0.061	−0.040	−0.005	0.000	0.000	0.006	0.001	0.001	0.000	−0.001	0
	0.25	0.010	0.014	0.012	0.009	0.009	0.007	0.004	0.001	0.000	−0.001	0
	0.30	0.039	0.034	0.017	0.011	0.011	0.006	0.004	0.001	0.000	−0.001	0
	0.40	0.040	0.032	0.014	0.008	0.008	0.004	0.003	0.001	0.000	0.000	0
	0.50	0.023	0.018	0.007	0.004	0.004	0.002	0.002	0.001	0.000	0.000	0
	0.60	0.011	0.008	0.003	0.002	0.002	0.001	0.001	0.000	0.000	0.000	0
	0.70	0.004	0.003	0.001	0.001	0.001	0.000	0.000	0.000	0.000	0.000	0
	0.80	0.000	0.000	0.000	0.000	0.000	0.000	0.000	0.000	0.000	0.000	0
$\bar{\tau}_{rz} / \eta_1$	0	0	0.209	0.472	$+\infty$	$+\infty$	0.439	0.398	0.386	0.387	0.390	0
	0.01	0	0.207	0.448	0.412	0.412	0.402	0.379	0.365	0.348	0.277	0
	0.02	0	0.200	0.393	0.347	0.347	0.342	0.332	0.310	0.263	0.142	0
	0.05	0	0.158	0.223	0.197	0.197	0.179	0.156	0.118	0.064	0.009	0
	0.10	0	0.061	0.034	0.014	0.014	0.004	−0.012	−0.026	−0.034	−0.028	0
	0.15	0	−0.014	−0.062	−0.066	−0.066	−0.064	−0.063	−0.059	−0.049	−0.029	0
	0.20	0	−0.052	−0.093	−0.086	−0.086	−0.079	−0.070	−0.059	−0.044	−0.025	0
	0.25	0	−0.062	−0.091	−0.079	−0.079	−0.071	−0.061	−0.050	−0.036	−0.019	0
	0.30	0	−0.057	−0.075	−0.064	−0.064	−0.056	−0.048	−0.038	−0.027	−0.014	0
	0.40	0	−0.034	−0.040	−0.033	−0.033	−0.029	−0.024	−0.019	−0.013	−0.007	0
	0.50	0	−0.015	−0.017	−0.014	−0.014	−0.012	−0.010	−0.008	−0.005	−0.003	0
	0.60	0	−0.006	−0.006	−0.005	−0.005	−0.004	−0.003	−0.003	−0.002	−0.001	0
	0.70	0	−0.002	−0.002	−0.001	−0.001	−0.001	−0.001	−0.001	0.000	0.000	0
	0.80	0	0.000	0.000	0.000	0.000	0.000	0.000	0.000	0.000	0.000	0

	λ	Core									Shell	
	p	0	0.3	0.6	0.7	0.75	0.80	0.85	0.90	0.95	0.95	1.00
$\dfrac{\bar{\sigma}_z}{\eta_1}$	0	0	0	0	0	0	0	0	0	0.168	0.003	0.6
	0.01	−0.040	−0.041	−0.047	−0.048	−0.047	−0.049	−0.044	−0.016	0.311	0.128	0.702
	0.02	−0.079	−0.082	−0.091	−0.092	−0.088	−0.081	−0.052	0.058	0.496	0.241	0.692
	0.05	−0.188	−0.193	−0.191	−0.163	−0.128	−0.056	0.077	0.323	0.722	0.362	0.602
	0.10	−0.319	−0.312	−0.225	−0.119	−0.032	0.085	0.240	0.434	0.656	0.328	0.444
	0.15	−0.371	−0.341	−0.174	−0.045	0.040	0.141	0.254	0.377	0.504	0.252	0.316
	0.20	−0.355	−0.308	−0.116	−0.002	0.065	0.138	0.213	0.289	0.363	0.181	0.217
	0.25	−0.300	−0.248	−0.071	0.017	0.066	0.113	0.161	0.207	0.248	0.124	0.144
	0.30	−0.232	−0.184	−0.041	0.022	0.053	0.085	0.114	0.140	0.163	0.081	0.091
	0.40	−0.114	−0.085	−0.011	0.016	0.028	0.039	0.049	0.056	0.060	0.030	0.032
	0.50	−0.045	−0.032	−0.001	0.008	0.012	0.015	0.017	0.018	0.017	0.009	0.008
	0.60	−0.014	−0.009	0.001	0.003	0.004	0.004	0.004	0.003	0.002	0.001	0.000
	0.70	−0.002	−0.001	0.001	0.001	0.001	0.001	0.000	0.000	−0.001	−0.001	−0.001
	0.80	0.000	0.000	0.000	0.000	0.000	0.000	0.000	0.000	0.000	0.000	0.000
$\dfrac{\bar{\sigma}_\theta}{\eta_1}$	0	−1.2	−1.2	−1.2	−1.2	−1.2	−1.2	−1.2	−1.2	−1.108	−0.635	−0.4
	0.01	−1.120	−1.115	−1.095	−1.080	−1.068	−1.051	−1.024	−0.971	−0.829	−0.443	−0.282
	0.02	−1.041	−1.031	−0.992	−0.962	−0.940	−0.908	−0.860	−0.780	−0.635	−0.326	−0.222
	0.05	−0.811	−0.790	−0.705	−0.646	−0.605	−0.554	−0.490	−0.409	−0.314	−0.157	−0.109
	0.10	−0.479	−0.449	−0.343	−0.282	−0.247	−0.207	−0.165	−0.122	−0.078	−0.039	−0.018
	0.15	−0.238	−0.211	−0.127	−0.087	−0.066	−0.044	−0.022	−0.001	0.020	0.010	0.020
	0.20	−0.085	−0.067	−0.016	0.006	0.016	0.027	0.037	0.047	0.056	0.028	0.032
	0.25	−0.002	0.008	0.033	0.043	0.047	0.052	0.055	0.059	0.062	0.031	0.032
	0.30	0.035	0.040	0.049	0.052	0.053	0.054	0.055	0.055	0.055	0.028	0.028
	0.40	0.044	0.043	0.040	0.038	0.037	0.036	0.035	0.034	0.032	0.016	0.016
	0.50	0.027	0.026	0.022	0.020	0.019	0.018	0.017	0.016	0.015	0.007	0.007
	0.60	0.013	0.012	0.010	0.008	0.008	0.007	0.007	0.006	0.006	0.003	0.003
	0.70	0.005	0.005	0.003	0.003	0.003	0.002	0.002	0.002	0.002	0.001	0.001
	0.80	0.000	0.000	0.000	0.000	0.000	0.000	0.000	0.000	0.000	0.000	0.000
$\dfrac{\bar{\sigma}_r}{\eta_1}$	0	−1.2	−1.2	−1.2	−1.2	−1.2	−1.2	−1.2	−1.2	−0.815	−0.815	0
	0.01	−1.120	−1.109	−1.059	−1.017	−0.983	−0.926	−0.833	−0.645	−0.287	−0.287	0
	0.02	−1.041	−1.020	−0.923	−0.842	−0.780	−0.682	−0.539	−0.321	−0.085	−0.085	0
	0.05	−0.811	−0.764	−0.568	−0.429	−0.336	−0.234	−0.125	−0.037	−0.001	−0.001	0
	0.10	−0.479	−0.417	−0.210	−0.116	−0.073	−0.033	−0.008	0.003	0.002	0.002	0
	0.15	−0.238	−0.188	−0.062	−0.023	−0.010	−0.001	0.002	0.002	0.000	0.000	0
	0.20	−0.085	−0.058	−0.006	0.003	0.005	0.004	0.002	0.000	−0.001	−0.001	0
	0.25	−0.002	0.007	0.014	0.010	0.006	0.004	0.001	−0.001	−0.001	−0.001	0
	0.30	0.035	0.033	0.019	0.011	0.008	0.003	0.001	−0.001	−0.001	−0.001	0
	0.40	0.044	0.036	0.015	0.007	0.004	0.002	0.000	−0.001	0.000	0.000	0
	0.50	0.027	0.021	0.008	0.004	0.002	0.001	0.000	0.000	0.000	0.000	0
	0.60	0.013	0.010	0.004	0.002	0.001	0.000	0.000	0.000	0.000	0.000	0
	0.70	0.005	0.004	0.001	0.001	0.001	0.000	0.000	0.000	0.000	0.000	0
	0.80	0.000	0.000	0.000	0.000	0.000	0.000	0.000	0.000	0.000	0.000	0
$\dfrac{\bar{\tau}_{rz}}{\eta_1}$	0	0	0.194	0.415	0.499	0.541	0.590	0.635	0.678	$+\infty$	$+\infty$	0
	0.01	0	0.192	0.409	0.488	0.526	0.565	0.589	0.565	0.323	0.323	0
	0.02	0	0.188	0.392	0.458	0.486	0.499	0.491	0.377	0.132	0.132	0
	0.05	0	0.158	0.288	0.294	0.275	0.230	0.154	0.055	−0.016	−0.016	0
	0.10	0	0.079	0.078	0.035	0.006	−0.027	−0.053	−0.060	−0.035	−0.035	0
	0.15	0	0.003	−0.053	−0.084	−0.096	−0.099	−0.092	−0.071	−0.031	−0.031	0
	0.20	0	−0.044	−0.104	−0.114	−0.112	−0.104	−0.088	−0.061	−0.024	−0.024	0
	0.25	0	−0.062	−0.109	−0.108	−0.101	−0.089	−0.071	−0.048	−0.018	−0.018	0
	0.30	0	−0.062	−0.094	−0.088	−0.080	−0.069	−0.054	−0.035	−0.013	−0.013	0
	0.40	0	−0.040	−0.053	−0.046	−0.041	−0.034	−0.026	−0.016	−0.006	−0.006	0
	0.50	0	−0.019	−0.023	−0.020	−0.017	−0.014	−0.010	−0.006	−0.002	−0.002	0
	0.60	0	−0.008	−0.008	−0.007	−0.005	−0.004	−0.003	−0.002	−0.001	−0.001	0
	0.70	0	−0.002	−0.002	−0.002	−0.001	−0.001	−0.001	0.000	0.000	0.000	0
	0.80	0	0.000	0.000	0.000	0.000	0.000	0.000	0.000	0.000	0.000	0

$y = 1,5; \mu = 0,7; \delta = 2$

	λ	Core				Shell						
	p	0	0.3	0.6	0.7	0.7	0.75	0.80	0.85	0.90	0.95	1.00
$\dfrac{\bar{\sigma}_z}{\eta_1}$	0	0	0	0	0.076	0.233	0	0	0	0	0	0.6
	0.01	−0.067	−0.073	−0.118	0.029	0.218	0.090	0.039	0.021	0.019	0.058	0.643
	0.02	−0.133	−0.144	−0.200	−0.013	0.214	0.126	0.071	0.050	0.067	0.192	0.673
	0.05	−0.315	−0.330	−0.289	−0.084	0.204	0.159	0.146	0.174	0.265	0.445	0.717
	0.10	−0.527	−0.512	−0.308	−0.110	0.219	0.224	0.257	0.319	0.411	0.532	0.674
	0.15	−0.607	−0.555	−0.285	−0.109	0.231	0.260	0.301	0.356	0.422	0.497	0.576
	0.20	−0.592	−0.520	−0.256	−0.118	0.231	0.264	0.301	0.342	0.387	0.432	0.477
	0.25	−0.527	−0.456	−0.234	−0.135	0.224	0.253	0.282	0.312	0.341	0.368	0.393
	0.30	−0.449	−0.388	−0.220	−0.152	0.216	0.238	0.260	0.280	0.298	0.315	0.328
	0.40	−0.319	−0.286	−0.206	−0.179	0.203	0.214	0.224	0.232	0.239	0.243	0.246
	0.50	−0.246	−0.232	−0.201	−0.192	0.196	0.200	0.204	0.207	0.209	0.210	0.210
	0.60	−0.214	−0.209	−0.200	−0.198	0.193	0.194	0.196	0.196	0.197	0.196	0.196
	0.70	−0.202	−0.201	−0.200	−0.199	0.192	0.192	0.192	0.192	0.192	0.192	0.192
	0.80	−0.200	−0.200	−0.200	−0.200	0.192	0.192	0.192	0.192	0.192	0.192	0.192
$\dfrac{\bar{\sigma}_\theta}{\eta_1}$	0	−1.8	−1.8	−1.8	−1.662	−0.635	−0.6	−0.6	−0.6	−0.6	−0.6	−0.4
	0.01	−1.680	−1.670	−1.612	−1.446	−0.517	−0.511	−0.507	−0.498	−0.480	−0.435	−0.285
	0.02	−1.560	−1.542	−1.434	−1.278	−0.421	−0.424	−0.418	−0.402	−0.371	−0.312	−0.205
	0.05	−1.218	−1.178	−0.999	−0.876	−0.191	−0.197	−0.188	−0.167	−0.135	−0.090	−0.040
	0.10	−0.739	−0.690	−0.524	−0.442	0.054	0.046	0.048	0.057	0.071	0.088	0.107
	0.15	−0.407	−0.369	−0.257	−0.208	0.182	0.170	0.164	0.162	0.163	0.166	0.171
	0.20	−0.207	−0.183	−0.120	−0.094	0.243	0.227	0.216	0.207	0.202	0.198	0.195
	0.25	−0.101	−0.088	−0.058	−0.047	0.268	0.250	0.235	0.224	0.214	0.206	0.200
	0.30	−0.053	−0.048	−0.037	−0.033	0.276	0.256	0.239	0.226	0.214	0.205	0.196
	0.40	−0.039	−0.040	−0.043	−0.045	0.270	0.249	0.231	0.216	0.203	0.192	0.183
	0.50	−0.056	−0.058	−0.062	−0.064	0.260	0.238	0.221	0.206	0.193	0.182	0.173
	0.60	−0.071	−0.072	−0.075	−0.076	0.254	0.232	0.214	0.199	0.187	0.176	0.167
	0.70	−0.080	−0.080	−0.081	−0.082	0.251	0.229	0.211	0.197	0.184	0.174	0.165
	0.80	−0.085	−0.085	−0.085	−0.085	0.249	0.228	0.210	0.195	0.183	0.173	0.164
$\dfrac{\bar{\sigma}_r}{\eta_1}$	0	−1.8	−1.8	−1.8	−1.045	−1.045	−0.6	−0.6	−0.6	−0.6	−0.6	0
	0.01	−1.680	−1.658	−1.471	−0.940	−0.940	−0.648	−0.538	−0.466	−0.379	−0.209	0
	0.02	−1.560	−1.518	−1.202	−0.823	−0.823	−0.614	−0.467	−0.351	−0.223	−0.067	0
	0.05	−1.218	−1.129	−0.725	−0.529	−0.529	−0.391	−0.267	−0.156	−0.066	−0.012	0
	0.10	−0.739	−0.638	−0.354	−0.253	−0.253	−0.176	−0.112	−0.063	−0.028	−0.009	0
	0.15	−0.407	−0.339	−0.188	−0.144	−0.144	−0.100	−0.066	−0.040	−0.022	−0.009	0
	0.20	−0.207	−0.174	−0.114	−0.100	−0.100	−0.072	−0.051	−0.034	−0.021	−0.010	0
	0.25	−0.101	−0.091	−0.082	−0.082	−0.082	−0.061	−0.045	−0.031	−0.020	−0.010	0
	0.30	−0.053	−0.056	−0.070	−0.076	−0.076	−0.057	−0.043	−0.030	−0.020	−0.010	0
	0.40	−0.039	−0.048	−0.069	−0.076	−0.076	−0.058	−0.043	−0.030	−0.019	−0.009	0
	0.50	−0.056	−0.063	−0.076	−0.080	−0.080	−0.060	−0.044	−0.031	−0.019	−0.009	0
	0.60	−0.071	−0.074	−0.081	−0.083	−0.083	−0.062	−0.045	−0.031	−0.019	−0.009	0
	0.70	−0.080	−0.081	−0.084	−0.084	−0.084	−0.063	−0.046	−0.031	−0.019	−0.009	0
	0.80	−0.085	−0.085	−0.085	−0.085	−0.085	−0.064	−0.046	−0.031	−0.019	−0.009	0
$\dfrac{\bar{\tau}_{rz}}{\eta_1}$	0	0	0.334	0.817	$+\infty$	$+\infty$	0.768	0.616	0.535	0.484	0.440	0
	0.01	0	0.331	0.777	0.829	0.829	0.708	0.591	0.512	0.443	0.325	0
	0.02	0	0.321	0.685	0.659	0.659	0.603	0.527	0.449	0.354	0.189	0
	0.05	0	0.260	0.403	0.372	0.372	0.338	0.288	0.219	0.132	0.044	0
	0.10	0	0.119	0.111	0.090	0.090	0.072	0.046	0.019	−0.004	−0.014	0
	0.15	0	0.008	−0.036	−0.040	−0.040	−0.041	−0.044	−0.045	−0.040	−0.026	0
	0.20	0	−0.050	−0.093	−0.084	−0.084	−0.077	−0.070	−0.060	−0.045	−0.026	0
	0.25	0	−0.070	−0.102	−0.088	−0.088	−0.078	−0.068	−0.055	−0.040	−0.022	0
	0.30	0	−0.067	−0.089	−0.075	−0.075	−0.066	−0.057	−0.045	−0.032	−0.017	0
	0.40	0	−0.042	−0.050	−0.041	−0.041	−0.036	−0.031	−0.024	−0.017	−0.009	0
	0.50	0	−0.020	−0.023	−0.019	−0.019	−0.016	−0.013	−0.010	−0.007	−0.004	0
	0.60	0	−0.008	−0.008	−0.007	−0.007	−0.006	−0.005	−0.004	−0.002	−0.001	0
	0.70	0	−0.002	−0.002	−0.002	−0.002	−0.002	−0.001	−0.001	−0.001	0.000	0
	0.80	0	0.000	0.000	0.000	0.000	0.000	0.000	0.000	0.000	0.000	0

	λ	Core					Shell					
	p	0	0.3	0.6	0.7	0.75	0.75	0.80	0.85	0.90	0.95	1.00
$\bar{\sigma}_z / \eta_1$	0	0	0	0	0	0.076	0.233	0	0	0	0	0.6
	0.01	−0.064	−0.069	−0.096	−0.140	0.375	0.229	0.096	0.045	0.034	0.070	0.659
	0.02	−0.128	−0.136	−0.177	−0.172	0.009	0.230	0.139	0.090	0.094	0.216	0.703
	0.05	−0.303	−0.316	−0.304	−0.167	−0.021	0.245	0.208	0.224	0.308	0.489	0.770
	0.10	−0.512	−0.500	−0.324	−0.131	0.006	0.283	0.309	0.366	0.458	0.581	0.726
	0.15	−0.595	−0.546	−0.282	−0.098	0.015	0.296	0.340	0.396	0.463	0.539	0.620
	0.20	−0.580	−0.509	−0.238	−0.090	−0.008	0.289	0.329	0.373	0.420	0.468	0.514
	0.25	−0.513	−0.440	−0.207	−0.098	−0.042	0.272	0.305	0.337	0.368	0.397	0.423
	0.30	−0.431	−0.367	−0.187	−0.112	−0.076	0.256	0.279	0.300	0.321	0.338	0.353
	0.40	−0.292	−0.255	−0.168	−0.139	−0.126	0.231	0.241	0.250	0.257	0.262	0.265
	0.50	−0.213	−0.197	−0.164	−0.154	−0.151	0.218	0.222	0.225	0.226	0.227	0.227
	0.60	−0.178	−0.173	−0.163	−0.161	−0.161	0.213	0.214	0.214	0.215	0.214	0.214
	0.70	−0.166	−0.165	−0.163	−0.163	−0.164	0.211	0.211	0.211	0.211	0.211	0.210
	0.80	−0.164	−0.164	−0.164	−0.164	−0.164	0.211	0.211	0.211	0.211	0.211	0.211
$\bar{\sigma}_\theta / \eta_1$	0	−1.8	−1.8	−1.8	−1.8	−1.662	−0.635	−0.6	−0.6	−0.6	−0.6	−0.4
	0.01	−1.681	−1.673	−1.629	−1.565	−1.441	−0.512	−0.502	−0.493	−0.475	−0.430	−0.279
	0.02	−1.563	−1.547	−1.464	−1.365	−1.264	−0.410	−0.407	−0.392	−0.362	−0.303	−0.195
	0.05	−1.224	−1.187	−1.030	−0.915	−0.843	−0.167	−0.165	−0.147	−0.116	−0.072	−0.020
	0.10	−0.744	−0.697	−0.533	−0.448	−0.401	0.080	0.078	0.084	0.097	0.113	0.131
	0.15	−0.405	−0.366	−0.251	−0.199	−0.172	0.203	0.195	0.191	0.190	0.192	0.196
	0.20	−0.197	−0.172	−0.105	−0.078	−0.064	0.260	0.246	0.236	0.229	0.224	0.220
	0.25	−0.086	−0.073	−0.040	−0.028	−0.022	0.282	0.265	0.252	0.241	0.232	0.224
	0.30	−0.036	−0.030	−0.018	−0.015	−0.013	0.287	0.269	0.253	0.240	0.229	0.220
	0.40	−0.022	−0.023	−0.026	−0.028	−0.029	0.279	0.258	0.242	0.228	0.216	0.205
	0.50	−0.041	−0.043	−0.047	−0.049	−0.050	0.268	0.248	0.231	0.217	0.204	0.194
	0.60	−0.057	−0.058	−0.061	−0.062	−0.063	0.261	0.241	0.225	0.210	0.198	0.188
	0.70	−0.066	−0.067	−0.068	−0.069	−0.069	0.258	0.238	0.222	0.208	0.196	0.186
	0.80	−0.072	−0.072	−0.072	−0.072	−0.072	0.257	0.237	0.221	0.207	0.195	0.185
$\bar{\sigma}_r / \eta_1$	0	−1.8	−1.8	−1.8	−1.8	−1.045	−1.045	−0.6	−0.6	−0.6	−0.6	0
	0.01	−1.681	−1.662	−1.534	−1.280	−0.911	−0.911	−0.616	−0.492	−0.388	−0.211	0
	0.02	−1.563	−1.526	−1.293	−0.985	−0.774	−0.774	−0.555	−0.389	−0.238	−0.071	0
	0.05	−1.224	−1.143	−0.776	−0.557	−0.450	−0.450	−0.308	−0.181	−0.078	−0.016	0
	0.10	−0.744	−0.647	−0.353	−0.244	−0.196	−0.196	−0.125	−0.071	−0.033	−0.010	0
	0.15	−0.405	−0.335	−0.174	−0.128	−0.111	−0.111	−0.074	−0.045	−0.025	−0.011	0
	0.20	−0.197	−0.161	−0.097	−0.084	−0.079	−0.079	−0.056	−0.038	−0.023	−0.011	0
	0.25	−0.086	−0.075	−0.065	−0.067	−0.068	−0.068	−0.050	−0.035	−0.022	−0.011	0
	0.30	−0.036	−0.039	−0.054	−0.061	−0.064	−0.064	−0.048	−0.034	−0.022	−0.011	0
	0.40	−0.022	−0.032	−0.055	−0.062	−0.065	−0.065	−0.048	−0.034	−0.022	−0.010	0
	0.50	−0.041	−0.048	−0.062	−0.066	−0.068	−0.068	−0.050	−0.035	−0.022	−0.010	0
	0.60	−0.057	−0.061	−0.068	−0.069	−0.070	−0.070	−0.051	−0.035	−0.022	−0.010	0
	0.70	−0.066	−0.068	−0.070	−0.071	−0.071	−0.071	−0.052	−0.036	−0.022	−0.010	0
	0.80	−0.072	−0.072	−0.072	−0.072	−0.072	−0.072	−0.052	−0.036	−0.022	−0.010	0
$\bar{\tau}_{rz} / \eta_1$	0	0	0.319	0.736	1.009	$+\infty$	$+\infty$	0.774	0.617	0.530	0.462	0
	0.01	0	0.316	0.715	0.889	0.829	0.829	0.710	0.584	0.486	0.347	0
	0.02	0	0.307	0.659	0.712	0.662	0.662	0.596	0.504	0.390	0.207	0
	0.05	0	0.253	0.422	0.379	0.349	0.349	0.303	0.234	0.143	0.050	0
	0.10	0	0.121	0.114	0.074	0.058	0.058	0.039	0.015	−0.007	−0.015	0
	0.15	0	0.010	−0.046	−0.061	−0.059	−0.059	−0.056	−0.053	−0.045	−0.029	0
	0.20	0	−0.052	−0.106	−0.103	−0.091	−0.091	−0.080	−0.067	−0.050	−0.028	0
	0.25	0	−0.074	−0.114	−0.102	−0.088	−0.088	−0.076	−0.061	−0.044	−0.024	0
	0.30	0	−0.072	−0.099	−0.085	−0.072	−0.072	−0.061	−0.049	−0.035	−0.018	0
	0.40	0	−0.046	−0.055	−0.045	−0.038	−0.038	−0.032	−0.025	−0.017	−0.009	0
	0.50	0	−0.021	−0.024	−0.019	−0.016	−0.016	−0.013	−0.010	−0.007	−0.004	0
	0.60	0	−0.008	−0.008	−0.007	−0.005	−0.005	−0.004	−0.003	−0.002	−0.001	0
	0.70	0	−0.002	−0.002	−0.002	−0.001	−0.001	−0.001	−0.001	−0.001	0.000	0
	0.80	0	0.000	0.000	0.000	0.000	0.000	0.000	0.000	0.000	0.000	0

$\gamma = 1,5 ; \mu = 0,8 ; \delta = 2$

	λ	Core						Shell				
	p	0	0.3	0.6	0.7	0.75	0.80	0.80	0.85	0.90	0.95	1.00
$\bar{\sigma}_z$ η_1^z	0	0	0	0	0	0	0.076	0.233	0	0	0	0.6
	0.01	−0.062	−0.066	−0.084	−0.106	−0.131	0.053	0.241	0.104	0.060	0.089	0.683
	0.02	−0.124	−0.131	−0.159	−0.176	−0.152	0.043	0.254	0.161	0.137	0.250	0.747
	0.05	−0.295	−0.305	−0.302	−0.219	−0.106	0.074	0.303	0.291	0.364	0.544	0.837
	0.10	−0.500	−0.489	−0.334	−0.153	−0.016	0.158	0.364	0.420	0.511	0.635	0.784
	0.15	−0.583	−0.537	−0.281	−0.093	0.025	0.159	0.372	0.433	0.504	0.583	0.667
	0.20	−0.569	−0.499	−0.224	−0.068	0.020	0.114	0.351	0.400	0.450	0.501	0.549
	0.25	−0.500	−0.425	−0.183	−0.067	−0.006	0.055	0.323	0.358	0.391	0.422	0.450
	0.30	−0.413	−0.347	−0.157	−0.077	−0.037	0.001	0.295	0.319	0.341	0.359	0.374
	0.40	−0.265	−0.227	−0.134	−0.101	−0.087	−0.075	0.257	0.267	0.274	0.279	0.282
	0.50	−0.181	−0.164	−0.128	−0.118	−0.114	−0.111	0.239	0.242	0.244	0.244	0.244
	0.60	−0.144	−0.138	−0.127	−0.125	−0.125	−0.125	0.231	0.232	0.232	0.232	0.231
	0.70	−0.131	−0.130	−0.128	−0.128	−0.129	−0.129	0.229	0.229	0.229	0.229	0.228
	0.80	−0.129	−0.129	−0.129	−0.129	−0.129	−0.129	0.229	0.229	0.229	0.229	0.230
$\bar{\sigma}_\theta$ η_1^θ	0	−1.8	−1.8	−1.8	−1.8	−1.8	−1.662	−0.635	−0.6	−0.6	−0.6	−0.4
	0.01	−1.682	−1.674	−1.638	−1.600	−1.556	−1.430	−0.502	−0.488	−0.469	−0.424	−0.272
	0.02	−1.565	−1.550	−1.480	−1.412	−1.347	−1.241	−0.391	−0.380	−0.351	−0.291	−0.182
	0.05	−1.227	−1.194	−1.051	−0.948	−0.878	−0.796	−0.134	−0.122	−0.093	−0.050	0.004
	0.10	−0.746	−0.700	−0.539	−0.452	−0.403	−0.350	0.111	0.115	0.125	0.140	0.158
	0.15	−0.401	−0.362	−0.243	−0.189	−0.160	−0.131	0.227	0.221	0.220	0.220	0.223
	0.20	−0.186	−0.160	−0.090	−0.061	−0.047	−0.032	0.278	0.266	0.258	0.251	0.246
	0.25	−0.070	−0.056	−0.022	−0.009	−0.003	0.003	0.296	0.281	0.269	0.258	0.249
	0.30	−0.018	−0.012	0.001	0.005	0.006	0.008	0.299	0.281	0.267	0.254	0.244
	0.40	−0.004	−0.005	−0.009	−0.011	−0.012	−0.014	0.288	0.269	0.253	0.239	0.227
	0.50	−0.025	−0.027	−0.032	−0.034	−0.035	−0.037	0.276	0.257	0.241	0.228	0.216
	0.60	−0.043	−0.044	−0.047	−0.048	−0.049	−0.050	0.269	0.251	0.235	0.222	0.210
	0.70	−0.053	−0.053	−0.055	−0.055	−0.055	−0.056	0.266	0.248	0.232	0.219	0.208
	0.80	−0.058	−0.058	−0.058	−0.058	−0.058	−0.058	0.265	0.247	0.231	0.218	0.207
$\bar{\sigma}_r$ η_1^r	0	−1.8	−1.8	−1.8	−1.8	−1.8	−1.045	−1.045	−0.6	−0.6	−0.6	0
	0.01	−1.682	−1.665	−1.565	−1.435	−1.252	−0.873	−0.873	−0.565	−0.409	−0.215	0
	0.02	−1.565	−1.531	−1.344	−1.133	−0.933	−0.706	−0.706	−0.468	−0.268	−0.079	0
	0.05	−1.227	−1.152	−0.818	−0.594	−0.471	−0.354	−0.354	−0.211	−0.094	−0.020	0
	0.10	−0.746	−0.651	−0.353	−0.233	−0.181	−0.138	−0.138	−0.078	−0.037	−0.012	0
	0.15	−0.401	−0.329	−0.159	−0.110	−0.092	−0.079	−0.079	−0.050	−0.028	−0.012	0
	0.20	−0.186	−0.148	−0.079	−0.066	−0.062	−0.061	−0.061	−0.041	−0.026	−0.012	0
	0.25	−0.070	−0.058	−0.047	−0.050	−0.052	−0.054	−0.054	−0.038	−0.025	−0.012	0
	0.30	−0.018	−0.021	−0.038	−0.046	−0.050	−0.053	−0.053	−0.038	−0.024	−0.012	0
	0.40	−0.004	−0.015	−0.040	−0.048	−0.051	−0.054	−0.054	−0.038	−0.024	−0.012	0
	0.50	−0.025	−0.032	−0.048	−0.053	−0.055	−0.056	−0.056	−0.039	−0.024	−0.011	0
	0.60	−0.043	−0.047	−0.054	−0.056	−0.057	−0.057	−0.057	−0.039	−0.024	−0.011	0
	0.70	−0.053	−0.054	−0.057	−0.057	−0.058	−0.058	−0.058	−0.040	−0.024	−0.011	0
	0.80	−0.058	−0.058	−0.058	−0.058	−0.058	−0.058	−0.058	−0.040	−0.024	−0.011	0
$\bar{\tau}_{rz}$ η_1^{rz}	0	0	0.307	0.684	0.872	1.028	$+\infty$	$+\infty$	0.768	0.602	0.496	0
	0.01	0	0.304	0.670	0.830	0.904	0.838	0.838	0.697	0.549	0.378	0
	0.02	0	0.297	0.631	0.729	0.721	0.647	0.647	0.566	0.435	0.231	0
	0.05	0	0.248	0.433	0.405	0.358	0.305	0.305	0.241	0.150	0.053	0
	0.10	0	0.122	0.119	0.068	0.040	0.019	0.019	0.003	−0.013	−0.018	0
	0.15	0	0.011	−0.053	−0.078	−0.081	−0.074	−0.074	−0.065	−0.053	−0.032	0
	0.20	0	−0.054	−0.118	−0.120	−0.111	−0.092	−0.092	−0.076	−0.057	−0.032	0
	0.25	0	−0.078	−0.126	−0.116	−0.103	−0.083	−0.083	−0.067	−0.048	−0.026	0
	0.30	0	−0.077	−0.109	−0.096	−0.083	−0.065	−0.065	−0.052	−0.037	−0.020	0
	0.40	0	−0.049	−0.060	−0.050	−0.042	−0.033	−0.033	−0.026	−0.018	−0.009	0
	0.50	0	−0.023	−0.026	−0.021	−0.017	−0.013	−0.013	−0.010	−0.007	−0.004	0
	0.60	0	−0.009	−0.009	−0.007	−0.005	−0.004	−0.004	−0.003	−0.002	−0.001	0
	0.70	0	−0.002	−0.002	−0.001	−0.001	−0.001	−0.001	−0.001	0.000	0.000	0
	0.80	0	0.000	0.000	0.000	0.000	0.000	0.000	0.000	0.000	0.000	0

	λ	Core							Shell			
	p	0	0.3	0.6	0.7	0.75	0.80	0.85	0.85	0.90	0.95	1.00
$\dfrac{\bar{\sigma}_z}{\eta_1}$	0	0	0	0	0	0	0	0.076	0.233	0	0	0.6
	0.01	−0.061	−0.064	−0.076	−0.087	−0.099	−0.117	0.086	0.254	0.121	0.120	0.722
	0.02	−0.121	−0.127	−0.147	−0.158	−0.159	−0.122	0.104	0.289	0.213	0.302	0.814
	0.05	−0.288	−0.297	−0.296	−0.240	−0.162	−0.014	0.230	0.392	0.440	0.616	0.919
	0.10	−0.491	−0.480	−0.341	−0.173	−0.040	0.136	0.356	0.470	0.565	0.692	0.845
	0.15	−0.574	−0.530	−0.281	−0.092	0.030	0.171	0.329	0.459	0.537	0.622	0.710
	0.20	−0.559	−0.490	−0.213	−0.052	0.041	0.141	0.244	0.418	0.472	0.527	0.578
	0.25	−0.488	−0.413	−0.163	−0.041	0.024	0.090	0.154	0.373	0.409	0.442	0.471
	0.30	−0.397	−0.329	−0.131	−0.046	−0.003	0.038	0.076	0.334	0.357	0.376	0.391
	0.40	−0.241	−0.201	−0.102	−0.067	−0.030	−0.038	−0.027	0.282	0.290	0.295	0.298
	0.50	−0.151	−0.133	−0.094	−0.083	−0.067	−0.075	−0.074	0.259	0.261	0.261	0.261
	0.60	−0.111	−0.105	−0.093	−0.091	−0.086	−0.091	−0.091	0.250	0.250	0.250	0.249
	0.70	−0.097	−0.096	−0.094	−0.094	−0.093	−0.095	−0.095	0.248	0.247	0.247	0.246
	0.80	−0.095	−0.095	−0.095	−0.095	−0.095	−0.095	−0.095	0.247	0.247	0.247	0.247
$\dfrac{\bar{\sigma}_\theta}{\eta_1}$	0	−1.8	−1.8	−1.8	−1.8	−1.8	−1.8	−1.662	−0.635	−0.6	−0.6	−0.4
	0.01	−1.632	−1.675	−1.643	−1.614	−1.590	−1.542	−1.410	−0.486	−0.462	−0.416	−0.261
	0.02	−1.566	−1.552	−1.489	−1.436	−1.391	−1.320	−1.202	−0.362	−0.336	−0.276	−0.163
	0.05	−1.228	−1.196	−1.064	−0.970	−0.905	−0.825	−0.730	−0.089	−0.065	−0.023	0.031
	0.10	−0.744	−0.699	−0.541	−0.453	−0.402	−0.347	−0.290	0.147	0.156	0.170	0.187
	0.15	−0.394	−0.355	−0.234	−0.177	−0.147	−0.116	−0.086	0.252	0.249	0.249	0.251
	0.20	−0.174	−0.147	−0.075	−0.044	−0.029	−0.014	0.001	0.296	0.286	0.278	0.273
	0.25	−0.054	−0.040	−0.004	0.010	0.016	0.023	0.028	0.310	0.296	0.284	0.274
	0.30	0.000	0.006	0.019	0.023	0.025	0.027	0.028	0.310	0.294	0.280	0.268
	0.40	0.013	0.012	0.008	0.006	0.004	0.003	0.001	0.297	0.279	0.264	0.251
	0.50	−0.009	−0.011	−0.016	−0.019	−0.020	−0.021	−0.023	0.284	0.267	0.252	0.239
	0.60	−0.028	−0.029	−0.032	−0.034	−0.035	−0.035	−0.036	0.278	0.260	0.245	0.233
	0.70	−0.038	−0.039	−0.040	−0.041	−0.041	−0.041	−0.042	0.275	0.257	0.243	0.230
	0.80	−0.044	−0.044	−0.044	−0.044	−0.044	−0.044	−0.044	0.274	0.256	0.242	0.229
$\dfrac{\bar{\sigma}_r}{\eta_1}$	0	−1.8	−1.8	−1.8	−1.8	−1.8	−1.8	−1.045	−1.045	−0.6	−0.6	0
	0.01	−1.682	−1.666	−1.583	−1.495	−1.404	−1.210	−0.818	−0.818	−0.472	−0.228	0
	0.02	−1.566	−1.534	−1.373	−1.217	−1.076	−0.857	−0.606	−0.606	−0.331	−0.095	0
	0.05	−1.228	−1.157	−0.848	−0.632	−0.499	−0.364	−0.241	−0.241	−0.109	−0.025	0
	0.10	−0.744	−0.651	−0.350	−0.221	−0.164	−0.116	−0.083	−0.083	−0.040	−0.013	0
	0.15	−0.394	−0.322	−0.143	−0.090	−0.071	−0.059	−0.052	−0.052	−0.030	−0.013	0
	0.20	−0.174	−0.134	−0.060	−0.047	−0.044	−0.043	−0.043	−0.043	−0.027	−0.013	0
	0.25	−0.054	−0.041	−0.029	−0.033	−0.036	−0.039	−0.041	−0.041	−0.027	−0.013	0
	0.30	0.000	−0.003	−0.021	−0.030	−0.035	−0.039	−0.041	−0.041	−0.027	−0.013	0
	0.40	0.013	0.002	−0.025	−0.035	−0.038	−0.040	−0.042	−0.042	−0.026	−0.013	0
	0.50	−0.009	−0.017	−0.034	−0.039	−0.041	−0.042	−0.043	−0.043	−0.027	−0.013	0
	0.60	−0.028	−0.032	−0.040	−0.042	−0.043	−0.043	−0.044	−0.044	−0.027	−0.012	0
	0.70	−0.038	−0.040	−0.043	−0.043	−0.044	−0.044	−0.044	−0.044	−0.027	−0.012	0
	0.80	−0.044	−0.044	−0.044	−0.044	−0.044	−0.044	−0.044	−0.044	−0.027	−0.012	0
$\dfrac{\bar{\tau}_{rz}}{\eta_1}$	0	0	0.298	0.650	0.802	0.899	1.041	$+\infty$	$+\infty$	0.740	0.550	0
	0.01	0	0.296	0.640	0.778	0.853	0.913	0.811	0.811	0.650	0.425	0
	0.02	0	0.289	0.609	0.713	0.744	0.718	0.607	0.607	0.486	0.261	0
	0.05	0	0.243	0.436	0.428	0.385	0.314	0.228	0.228	0.148	0.052	0
	0.10	0	0.123	0.125	0.067	0.031	−0.002	−0.020	−0.020	−0.027	−0.025	0
	0.15	0	0.011	−0.058	−0.091	−0.100	−0.098	−0.082	−0.082	−0.064	−0.038	0
	0.20	0	−0.056	−0.129	−0.136	−0.129	−0.113	−0.085	−0.085	−0.063	−0.035	0
	0.25	0	−0.082	−0.137	−0.130	−0.118	−0.099	−0.071	−0.071	−0.052	−0.028	0
	0.30	0	−0.082	−0.118	−0.107	−0.094	−0.076	−0.054	−0.054	−0.038	−0.020	0
	0.40	0	−0.052	−0.061	−0.055	−0.047	−0.037	−0.025	−0.025	−0.018	−0.009	0
	0.50	0	−0.025	−0.028	−0.023	−0.019	−0.015	−0.010	−0.010	−0.007	−0.003	0
	0.60	0	−0.009	−0.010	−0.007	−0.006	−0.004	−0.003	−0.003	−0.002	−0.001	0
	0.70	0	−0.003	−0.002	−0.002	−0.001	−0.001	−0.001	−0.001	−0.001	0.000	0
	0.80	0	0.000	0.000	0.000	0.000	0.000	0.000	0.000	0.000	0.000	0

γ = 1,5; μ = 0,9; δ = 2

	λ / p	Core								Shell		
		0	0.3	0.6	0.7	0.75	0.80	0.85	0.90	0.90	0.95	1.00
$\dfrac{\bar{\sigma}_z}{\eta_1}$	0	0	0	0	0	0	0	0	0.076	0.233	0	0.6
	0.01	−0.060	−0.063	−0.072	−0.076	−0.082	−0.088	−0.096	0.130	0.303	0.188	0.796
	0.02	−0.119	−0.124	−0.140	−0.143	−0.145	−0.135	−0.071	0.226	0.368	0.393	0.923
	0.05	−0.284	−0.292	−0.290	−0.249	−0.189	−0.075	0.139	0.500	0.536	0.705	1.015
	0.10	−0.485	−0.474	−0.345	−0.188	−0.061	0.112	0.335	0.611	0.601	0.742	0.902
	0.15	−0.568	−0.524	−0.282	−0.094	0.030	0.174	0.338	0.516	0.554	0.647	0.739
	0.20	−0.551	−0.483	−0.205	−0.041	0.056	0.159	0.268	0.377	0.485	0.543	0.595
	0.25	−0.477	−0.401	−0.147	−0.021	0.047	0.116	0.184	0.249	0.421	0.456	0.485
	0.30	−0.383	−0.314	−0.109	−0.020	0.025	0.069	0.110	0.147	0.370	0.389	0.405
	0.40	−0.219	−0.177	−0.073	−0.035	−0.019	−0.004	0.008	0.017	0.305	0.311	0.314
	0.50	−0.123	−0.104	−0.062	−0.050	−0.045	−0.041	−0.039	−0.038	0.278	0.278	0.278
	0.60	−0.080	−0.074	−0.060	−0.058	−0.057	−0.057	−0.058	−0.059	0.267	0.267	0.266
	0.70	−0.065	−0.063	−0.061	−0.061	−0.061	−0.062	−0.062	−0.062	0.265	0.265	0.264
	0.80	−0.062	−0.062	−0.062	−0.062	−0.062	−0.062	−0.062	−0.062	0.265	0.265	0.265
$\dfrac{\bar{\sigma}_\theta}{\eta_1}$	0	−1.8	−1.8	−1.8	−1.8	−1.8	−1.8	−1.8	−1.662	−0.635	−0.6	−0.4
	0.01	−1.682	−1.675	−1.645	−1.622	−1.602	−1.573	−1.518	−1.371	−0.455	−0.405	−0.243
	0.02	−1.566	−1.552	−1.493	−1.448	−1.412	−1.359	−1.275	−1.134	−0.311	−0.255	−0.136
	0.05	−1.227	−1.196	−1.070	−0.982	−0.921	−0.845	−0.750	−0.637	−0.031	0.009	0.063
	0.10	−0.741	−0.696	−0.539	−0.451	−0.399	−0.343	−0.283	−0.220	0.186	0.200	0.216
	0.15	−0.386	−0.346	−0.224	−0.166	−0.134	−0.102	−0.070	−0.039	0.278	0.277	0.277
	0.20	−0.162	−0.135	−0.060	−0.028	−0.012	0.003	0.019	0.033	0.314	0.305	0.297
	0.25	−0.039	−0.024	0.013	0.027	0.034	0.041	0.047	0.052	0.323	0.310	0.298
	0.30	0.017	0.023	0.037	0.041	0.043	0.045	0.046	0.047	0.320	0.305	0.292
	0.40	0.031	0.030	0.025	0.023	0.022	0.020	0.018	0.017	0.305	0.289	0.274
	0.50	0.008	0.006	0.000	−0.002	−0.004	−0.005	−0.007	−0.008	0.293	0.276	0.262
	0.60	−0.012	−0.013	−0.017	−0.018	−0.019	−0.020	−0.021	−0.022	0.286	0.270	0.256
	0.70	−0.023	−0.024	−0.025	−0.026	−0.026	−0.027	−0.027	−0.027	0.283	0.267	0.254
	0.80	−0.030	−0.030	−0.030	−0.030	−0.030	−0.030	−0.030	−0.030	0.282	0.266	0.253
$\dfrac{\bar{\sigma}_r}{\eta_1}$	0	−1.8	−1.8	−1.8	−1.8	−1.8	−1.8	−1.8	−1.045	−1.045	−0.6	0
	0.01	−1.682	−1.667	−1.591	−1.524	−1.460	−1.356	−1.139	−0.694	−0.694	−0.271	0
	0.02	−1.566	−1.535	−1.388	−1.262	−1.152	−0.989	−0.736	−0.437	−0.437	−0.129	0
	0.05	−1.227	−1.159	−0.865	−0.657	−0.526	−0.378	−0.233	−0.122	−0.122	−0.028	0
	0.10	−0.741	−0.649	−0.346	−0.211	−0.146	−0.093	−0.057	−0.040	−0.040	−0.014	0
	0.15	−0.386	−0.313	−0.128	−0.072	−0.051	−0.037	−0.031	−0.030	−0.030	−0.014	0
	0.20	−0.162	−0.121	−0.043	−0.028	−0.026	−0.026	−0.027	−0.029	−0.029	−0.014	0
	0.25	−0.039	−0.025	−0.013	−0.018	−0.021	−0.024	−0.028	−0.030	−0.030	−0.015	0
	0.30	0.017	0.014	−0.004	−0.015	−0.020	−0.025	−0.028	−0.030	−0.030	−0.015	0
	0.40	0.031	0.019	−0.010	−0.020	−0.023	−0.027	−0.029	−0.030	−0.030	−0.014	0
	0.50	0.008	0.000	−0.019	−0.025	−0.027	−0.028	−0.030	−0.030	−0.030	−0.014	0
	0.60	−0.012	−0.016	−0.025	−0.027	−0.028	−0.029	−0.030	−0.030	−0.030	−0.014	0
	0.70	−0.023	−0.025	−0.028	−0.029	−0.029	−0.029	−0.030	−0.030	−0.030	−0.014	0
	0.80	−0.030	−0.030	−0.030	−0.030	−0.030	−0.030	−0.030	−0.030	−0.030	−0.014	0
$\dfrac{\bar{\tau}_{rz}}{\eta_1}$	0	0	0.293	0.630	0.762	0.840	0.927	1.049	$+\infty$	$+\infty$	0.664	0
	0.01	0	0.291	0.621	0.745	0.812	0.875	0.911	0.749	0.749	0.501	0
	0.02	0	0.284	0.594	0.697	0.737	0.751	0.690	0.512	0.512	0.289	0
	0.05	0	0.240	0.438	0.443	0.411	0.342	0.236	0.118	0.118	0.040	0
	0.10	0	0.123	0.129	0.071	0.028	−0.013	−0.045	−0.050	−0.050	−0.036	0
	0.15	0	0.011	−0.063	−0.103	−0.115	−0.119	−0.107	−0.073	−0.073	−0.043	0
	0.20	0	−0.058	−0.138	−0.150	−0.146	−0.132	−0.106	−0.066	−0.066	−0.037	0
	0.25	0	−0.086	−0.148	−0.143	−0.132	−0.114	−0.088	−0.052	−0.052	−0.028	0
	0.30	0	−0.086	−0.128	−0.118	−0.105	−0.088	−0.065	−0.037	−0.037	−0.020	0
	0.40	0	−0.056	−0.071	−0.061	−0.053	−0.043	−0.031	−0.017	−0.017	−0.009	0
	0.50	0	−0.027	−0.031	−0.025	−0.022	−0.017	−0.012	−0.007	−0.007	−0.003	0
	0.60	0	−0.010	−0.010	−0.008	−0.007	−0.005	−0.003	−0.001	−0.001	−0.001	0
	0.70	0	−0.003	−0.003	−0.002	−0.002	−0.001	−0.001	0.000	0.000	0.000	0
	0.80	0	0.000	0.000	0.000	0.000	0.000	0.000	0.000	0.000	0.000	0

	λ	Core									Shell	
	p	0	0.3	0.6	0.7	0.75	0.80	0.85	0.90	0.95	0.95	1.00
$\dfrac{\bar{\sigma}_z}{\eta_1}$	0	0	0	0	0	0	0	0	0	0.076	0.233	0.6
	0.01	−0.059	−0.062	−0.070	−0.073	−0.072	−0.076	−0.074	−0.055	0.309	0.414	0.967
	0.02	−0.118	−0.123	−0.137	−0.140	−0.135	−0.130	−0.097	0.035	0.580	0.577	1.099
	0.05	−0.281	−0.288	−0.289	−0.253	−0.207	−0.110	0.070	0.403	0.954	0.776	1.110
	0.10	−0.480	−0.470	−0.350	−0.202	−0.080	0.085	0.306	0.584	0.908	0.752	0.923
	0.15	−0.562	−0.520	−0.284	−0.099	0.024	0.169	0.334	0.515	0.702	0.650	0.744
	0.20	−0.545	−0.477	−0.201	−0.035	0.062	0.168	0.279	0.392	0.501	0.549	0.602
	0.25	−0.468	−0.392	−0.136	−0.007	0.064	0.134	0.205	0.274	0.335	0.466	0.496
	0.30	−0.371	−0.301	−0.092	0.000	0.047	0.094	0.137	0.177	0.210	0.403	0.419
	0.40	−0.199	−0.156	−0.047	−0.007	0.011	0.028	0.041	0.053	0.060	0.328	0.351
	0.50	−0.098	−0.078	−0.033	−0.019	−0.012	−0.008	−0.005	−0.004	−0.004	0.296	0.296
	0.60	−0.052	−0.044	−0.029	−0.026	−0.024	−0.024	−0.024	−0.025	−0.027	0.285	0.284
	0.70	−0.034	−0.032	−0.029	−0.029	−0.030	−0.030	−0.030	−0.031	−0.031	0.282	0.282
	0.80	−0.031	−0.031	−0.031	−0.031	−0.031	−0.031	−0.031	−0.031	−0.031	0.283	0.283
$\dfrac{\bar{\sigma}_\theta}{\eta_1}$	0	−1.8	−1.8	−1.8	−1.8	−1.8	−1.8	−1.8	−1.8	−1.662	−0.635	−0.4
	0.01	−1.682	−1.675	−1.647	−1.624	−1.608	−1.583	−1.545	−1.472	−1.276	−0.381	−0.206
	0.02	−1.565	−1.552	−1.495	−1.453	−1.421	−1.376	−1.308	−1.196	−0.997	−0.215	−0.094
	0.05	−1.226	−1.196	−1.073	−0.988	−0.930	−0.857	−0.764	−0.649	−0.512	0.041	0.096
	0.10	−0.737	−0.693	−0.537	−0.449	−0.397	−0.340	−0.278	−0.213	−0.148	0.225	0.241
	0.15	−0.379	−0.339	−0.216	−0.157	−0.125	−0.092	−0.059	−0.027	0.004	0.301	0.300
	0.20	−0.151	−0.123	−0.047	−0.015	0.001	0.017	0.033	0.048	0.061	0.329	0.321
	0.25	−0.025	−0.010	0.029	0.043	0.050	0.057	0.063	0.069	0.073	0.335	0.322
	0.30	0.033	0.039	0.054	0.059	0.061	0.062	0.063	0.064	0.065	0.331	0.316
	0.40	0.049	0.048	0.043	0.041	0.039	0.038	0.036	0.034	0.033	0.315	0.299
	0.50	0.025	0.023	0.017	0.015	0.013	0.012	0.010	0.009	0.007	0.302	0.287
	0.60	0.004	0.003	−0.001	−0.002	−0.003	−0.004	−0.005	−0.006	−0.007	0.295	0.280
	0.70	−0.007	−0.008	−0.010	−0.011	−0.011	−0.011	−0.012	−0.012	−0.012	0.292	0.277
	0.80	−0.015	−0.015	−0.015	−0.015	−0.015	−0.015	−0.015	−0.015	−0.015	0.291	0.276
$\dfrac{\bar{\sigma}_r}{\eta_1}$	0	−1.8	−1.8	−1.8	−1.8	−1.8	−1.8	−1.8	−1.8	−1.045	−1.045	0
	0.01	−1.682	−1.667	−1.595	−1.534	−1.486	−1.404	−1.271	−0.996	−0.430	−0.430	0
	0.02	−1.565	−1.535	−1.395	−1.279	−1.191	−1.051	−0.845	−0.527	−0.166	−0.166	0
	0.05	−1.226	−1.159	−0.874	−0.673	−0.538	−0.389	−0.227	−0.090	−0.023	−0.023	0
	0.10	−0.737	−0.646	−0.342	−0.201	−0.135	−0.075	−0.034	−0.014	−0.013	−0.013	0
	0.15	−0.379	−0.305	−0.116	−0.057	−0.034	−0.020	−0.014	−0.013	−0.014	−0.014	0
	0.20	−0.151	−0.109	−0.028	−0.013	−0.010	−0.011	−0.013	−0.015	−0.016	−0.016	0
	0.25	−0.025	−0.010	−0.003	−0.002	−0.007	−0.010	−0.014	−0.016	−0.016	−0.016	0
	0.30	0.033	0.031	0.012	0.001	−0.004	−0.010	−0.014	−0.016	−0.016	−0.016	0
	0.40	0.049	0.037	0.007	−0.004	−0.008	−0.011	−0.015	−0.016	−0.016	−0.016	0
	0.50	0.025	0.017	−0.003	−0.009	−0.012	−0.013	−0.015	−0.015	−0.015	−0.015	0
	0.60	0.004	0.000	−0.010	−0.012	−0.014	−0.014	−0.015	−0.015	−0.015	−0.015	0
	0.70	−0.007	−0.009	−0.013	−0.014	−0.014	−0.015	−0.015	−0.015	−0.015	−0.015	0
	0.80	−0.015	−0.015	−0.015	−0.015	−0.015	−0.015	−0.015	−0.015	−0.015	−0.015	0
$\dfrac{\bar{\tau}_{rz}}{\eta_1}$	0	0	0.290	0.621	0.747	0.812	0.888	0.964	1.055	$+\infty$	$+\infty$	0
	0.01	0	0.287	0.612	0.732	0.791	0.852	0.898	0.886	0.567	0.576	0
	0.02	0	0.281	0.587	0.689	0.733	0.758	0.741	0.608	0.261	0.261	0
	0.05	0	0.238	0.438	0.452	0.429	0.368	0.263	0.119	0.002	0.002	0
	0.10	0	0.122	0.131	0.071	0.029	−0.020	−0.060	−0.075	−0.046	−0.046	0
	0.15	0	0.010	−0.067	−0.111	−0.129	−0.135	−0.127	−0.099	−0.044	−0.044	0
	0.20	0	−0.060	−0.146	−0.161	−0.158	−0.149	−0.125	−0.088	−0.035	−0.035	0
	0.25	0	−0.089	−0.156	−0.155	−0.146	−0.129	−0.104	−0.070	−0.027	−0.027	0
	0.30	0	−0.089	−0.136	−0.128	−0.117	−0.100	−0.079	−0.051	−0.019	−0.019	0
	0.40	0	−0.059	−0.077	−0.068	−0.061	−0.050	−0.038	−0.024	−0.008	−0.008	0
	0.50	0	−0.029	−0.035	−0.029	−0.025	−0.021	−0.015	−0.009	−0.003	−0.003	0
	0.60	0	−0.011	−0.012	−0.010	−0.008	−0.007	−0.005	−0.003	−0.001	−0.001	0
	0.70	0	−0.004	−0.003	−0.003	−0.002	−0.001	−0.001	−0.001	0.000	0.000	0
	0.80	0	0.000	0.000	0.000	0.000	0.000	0.000	0.000	0.000	0.000	0

$\gamma = 2; \mu = 0,7; \delta = 2$

	λ	Core				Shell						
	p	0	0.3	0.6	0.7	0.7	0.75	0.80	0.85	0.90	0.95	1.00
$\bar{\sigma}_z / \eta_1$	0	0	0	0	-0.016	0.464	0	0	0	0	0	0.6
	0.01	-0.092	-0.101	-0.178	-0.079	0.440	0.167	0.081	0.051	0.046	0.086	0.676
	0.02	-0.182	-0.199	-0.304	-0.135	0.433	0.248	0.148	0.109	0.121	0.247	0.737
	0.05	-0.432	-0.458	-0.452	-0.227	0.412	0.328	0.286	0.298	0.384	0.571	0.863
	0.10	-0.729	-0.720	-0.501	-0.274	0.422	0.414	0.438	0.498	0.595	0.730	0.895
	0.15	-0.856	-0.798	-0.488	-0.283	0.432	0.458	0.501	0.560	0.634	0.722	0.819
	0.20	-0.855	-0.772	-0.461	-0.298	0.431	0.466	0.507	0.554	0.607	0.662	0.719
	0.25	-0.789	-0.703	-0.439	-0.319	0.423	0.455	0.489	0.524	0.559	0.594	0.626
	0.30	-0.701	-0.627	-0.423	-0.340	0.413	0.439	0.465	0.489	0.512	0.533	0.550
	0.40	-0.548	-0.506	-0.407	-0.373	0.397	0.411	0.423	0.433	0.442	0.448	0.452
	0.50	-0.459	-0.440	-0.401	-0.389	0.389	0.394	0.399	0.403	0.405	0.407	0.407
	0.60	-0.418	-0.412	-0.399	-0.396	0.385	0.387	0.388	0.389	0.390	0.390	0.389
	0.70	-0.403	-0.401	-0.399	-0.398	0.384	0.384	0.384	0.385	0.385	0.384	0.384
	0.80	-0.399	-0.399	-0.399	-0.399	0.383	0.383	0.383	0.383	0.383	0.383	0.383
$\bar{\sigma}_\theta / \eta_1$	0	-2.4	-2.4	-2.4	-2.215	-0.635	-0.6	-0.6	-0.6	-0.6	-0.6	-0.4
	0.01	-2.246	-2.234	-2.157	-1.933	-0.483	-0.489	-0.489	-0.481	-0.464	-0.420	-0.269
	0.02	-2.093	-2.070	-1.928	-1.717	-0.363	-0.380	-0.381	-0.369	-0.340	-0.281	-0.174
	0.05	-1.655	-1.604	-1.371	-1.210	-0.077	-0.101	-0.103	-0.090	-0.061	-0.019	0.032
	0.10	-1.042	-0.978	-0.766	-0.661	0.229	0.203	0.191	0.190	0.198	0.211	0.228
	0.15	-0.614	-0.565	-0.422	-0.359	0.394	0.363	0.342	0.329	0.321	0.318	0.318
	0.20	-0.353	-0.322	-0.240	-0.206	0.477	0.441	0.413	0.393	0.377	0.364	0.355
	0.25	-0.211	-0.195	-0.154	-0.138	0.514	0.474	0.443	0.418	0.397	0.380	0.366
	0.30	-0.145	-0.138	-0.121	-0.116	0.526	0.485	0.451	0.424	0.401	0.380	0.364
	0.40	-0.119	-0.119	-0.122	-0.123	0.522	0.480	0.444	0.415	0.390	0.369	0.350
	0.50	-0.136	-0.138	-0.142	-0.144	0.511	0.468	0.433	0.403	0.378	0.357	0.338
	0.60	-0.153	-0.155	-0.158	-0.159	0.504	0.461	0.425	0.396	0.371	0.350	0.332
	0.70	-0.164	-0.164	-0.166	-0.166	0.500	0.457	0.421	0.392	0.367	0.347	0.329
	0.80	-0.170	-0.170	-0.170	-0.170	0.498	0.455	0.420	0.390	0.366	0.345	0.327
$\bar{\sigma}_r / \eta_1$	0	-2.4	-2.4	-2.4	-1.276	-1.275	-0.6	-0.6	-0.6	-0.6	-0.6	0
	0.01	-2.246	-2.217	-1.962	-1.170	-1.170	-0.719	-0.568	-0.481	-0.385	-0.211	0
	0.02	-2.093	-2.037	-1.607	-1.043	-1.043	-0.723	-0.521	-0.378	-0.236	-0.071	0
	0.05	-1.655	-1.538	-0.996	-0.723	-0.723	-0.521	-0.348	-0.202	-0.089	-0.019	0
	0.10	-1.042	-0.910	-0.535	-0.401	-0.401	-0.282	-0.184	-0.108	-0.052	-0.018	0
	0.15	-0.614	-0.524	-0.322	-0.261	-0.261	-0.185	-0.125	-0.078	-0.043	-0.018	0
	0.20	-0.353	-0.308	-0.222	-0.199	-0.199	-0.146	-0.103	-0.068	-0.041	-0.019	0
	0.25	-0.211	-0.196	-0.176	-0.173	-0.173	-0.129	-0.093	-0.064	-0.040	-0.019	0
	0.30	-0.145	-0.146	-0.157	-0.162	-0.162	-0.122	-0.089	-0.062	-0.039	-0.019	0
	0.40	-0.119	-0.129	-0.153	-0.160	-0.160	-0.120	-0.088	-0.061	-0.038	-0.018	0
	0.50	-0.136	-0.143	-0.159	-0.164	-0.164	-0.123	-0.090	-0.062	-0.038	-0.018	0
	0.60	-0.153	-0.157	-0.165	-0.167	-0.167	-0.125	-0.091	-0.062	-0.038	-0.018	0
	0.70	-0.164	-0.165	-0.168	-0.169	-0.169	-0.126	-0.092	-0.063	-0.038	-0.018	0
	0.80	-0.170	-0.170	-0.170	-0.170	-0.170	-0.127	-0.092	-0.063	-0.038	-0.018	0
$\bar{\tau}^{rz} / \eta_1$	0	0	0.459	1.162	$+\infty$	$+\infty$	1.097	0.834	0.686	0.581	0.489	0
	0.01	0	0.455	1.106	1.246	1.246	1.013	0.802	0.659	0.539	0.374	0
	0.02	0	0.442	0.977	0.971	0.971	0.865	0.722	0.588	0.446	0.235	0
	0.05	0	0.361	0.583	0.547	0.547	0.497	0.421	0.320	0.201	0.079	0
	0.10	0	0.176	0.188	0.165	0.165	0.140	0.104	0.063	0.026	0.001	0
	0.15	0	0.031	-0.010	-0.015	-0.015	-0.018	-0.025	-0.031	-0.032	-0.023	0
	0.20	0	-0.048	-0.093	-0.082	-0.082	-0.076	-0.069	-0.060	-0.047	-0.027	0
	0.25	0	-0.077	-0.113	-0.096	-0.096	-0.086	-0.075	-0.061	-0.045	-0.025	0
	0.30	0	-0.078	-0.103	-0.086	-0.086	-0.076	-0.065	-0.052	-0.037	-0.020	0
	0.40	0	-0.051	-0.061	-0.050	-0.050	-0.044	-0.037	-0.029	-0.020	-0.010	0
	0.50	0	-0.025	-0.028	-0.023	-0.023	-0.020	-0.017	-0.013	-0.009	-0.004	0
	0.60	0	-0.010	-0.011	-0.009	-0.009	-0.008	-0.006	-0.005	-0.003	-0.002	0
	0.70	0	-0.003	-0.003	-0.003	-0.003	-0.002	-0.002	-0.001	-0.001	0.000	0
	0.80	0	0.000	0.000	0.000	0.000	0.000	0.000	0.000	0.000	0.000	0

	λ	Core					Shell					
	p	0	0.3	0.6	0.7	0.75	0.75	0.80	0.85	0.90	0.95	1.00
$\dfrac{\bar{\sigma}_z}{\eta_1}$	0	0	0	0	0	−0.016	0.464	0	0	0	0	0.6
	0.01	−0.087	−0.094	−0.140	−0.226	−0.067	0.455	0.175	0.091	0.071	0.107	0.703
	0.02	−0.173	−0.186	−0.258	−0.290	−0.103	0.455	0.267	0.175	0.165	0.288	0.789
	0.05	−0.412	−0.433	−0.450	−0.306	−0.143	0.468	0.393	0.384	0.459	0.649	0.956
	0.10	−0.701	−0.693	−0.501	−0.278	−0.119	0.508	0.525	0.581	0.680	0.819	0.993
	0.15	−0.827	−0.772	−0.464	−0.246	−0.111	0.524	0.570	0.632	0.710	0.802	0.903
	0.20	−0.825	−0.742	−0.417	−0.238	−0.137	0.516	0.562	0.614	0.671	0.730	0.789
	0.25	−0.753	−0.665	−0.381	−0.247	−0.177	0.497	0.536	0.575	0.613	0.650	0.685
	0.30	−0.658	−0.579	−0.357	−0.264	−0.218	0.478	0.506	0.533	0.558	0.581	0.600
	0.40	−0.489	−0.444	−0.334	−0.296	−0.279	0.447	0.460	0.471	0.481	0.488	0.492
	0.50	−0.391	−0.371	−0.328	−0.315	−0.310	0.431	0.436	0.440	0.442	0.444	0.444
	0.60	−0.347	−0.340	−0.327	−0.324	−0.323	0.424	0.426	0.426	0.427	0.427	0.426
	0.70	−0.331	−0.330	−0.327	−0.327	−0.328	0.422	0.422	0.422	0.422	0.422	0.421
	0.80	−0.328	−0.328	−0.328	−0.328	−0.328	0.422	0.422	0.422	0.422	0.422	0.422
$\dfrac{\bar{\sigma}_\theta}{\eta_1}$	0	−2.4	−2.4	−2.4	−2.4	−2.215	−0.635	−0.6	−0.6	−0.6	−0.6	−0.4
	0.01	−2.247	−2.237	−2.180	−2.094	−1.927	−0.477	−0.478	−0.472	−0.456	−0.411	−0.260
	0.02	−2.096	−2.075	−1.967	−1.836	−1.701	−0.350	−0.359	−0.351	−0.324	−0.265	−0.156
	0.05	−1.660	−1.614	−1.410	−1.262	−1.169	−0.047	−0.060	−0.052	−0.026	0.016	0.069
	0.10	−1.042	−0.982	−0.772	−0.664	−0.604	0.266	0.248	0.243	0.247	0.259	0.275
	0.15	−0.603	−0.553	−0.405	−0.338	−0.303	0.428	0.402	0.385	0.375	0.369	0.367
	0.20	−0.330	−0.298	−0.210	−0.174	−0.156	0.506	0.474	0.450	0.431	0.416	0.404
	0.25	−0.181	−0.163	−0.119	−0.103	−0.095	0.538	0.503	0.474	0.450	0.431	0.414
	0.30	−0.111	−0.103	−0.086	−0.080	−0.077	0.548	0.510	0.479	0.452	0.430	0.411
	0.40	−0.085	−0.086	−0.089	−0.091	−0.092	0.540	0.501	0.467	0.439	0.415	0.395
	0.50	−0.106	−0.108	−0.113	−0.115	−0.117	0.528	0.488	0.454	0.426	0.402	0.381
	0.60	−0.126	−0.127	−0.130	−0.132	−0.133	0.520	0.480	0.446	0.418	0.394	0.374
	0.70	−0.137	−0.137	−0.139	−0.140	−0.140	0.516	0.476	0.443	0.415	0.391	0.371
	0.80	−0.144	−0.144	−0.144	−0.144	−0.144	0.514	0.474	0.441	0.414	0.390	0.370
$\dfrac{\bar{\sigma}_r}{\eta_1}$	0	−2.4	−2.4	−2.4	−2.4	−1.276	−1.276	−0.6	−0.6	−0.6	−0.6	0
	0.01	−2.247	−2.222	−2.053	−1.694	−1.136	−1.136	−0.683	−0.518	−0.399	−0.214	0
	0.02	−2.096	−2.047	−1.737	−1.304	−0.985	−0.985	−0.657	−0.436	−0.258	−0.077	0
	0.05	−1.660	−1.555	−1.069	−0.772	−0.625	−0.625	−0.422	−0.247	−0.110	−0.026	0
	0.10	−1.042	−0.916	−0.531	−0.385	−0.320	−0.320	−0.211	−0.124	−0.061	−0.021	0
	0.15	−0.603	−0.511	−0.296	−0.232	−0.206	−0.206	−0.140	−0.089	−0.050	−0.021	0
	0.20	−0.330	−0.282	−0.190	−0.169	−0.161	−0.161	−0.114	−0.077	−0.046	−0.021	0
	0.25	−0.181	−0.164	−0.144	−0.143	−0.143	−0.143	−0.104	−0.072	−0.045	−0.021	0
	0.30	−0.111	−0.112	−0.126	−0.133	−0.136	−0.136	−0.100	−0.070	−0.044	−0.021	0
	0.40	−0.085	−0.097	−0.124	−0.132	−0.135	−0.135	−0.099	−0.069	−0.043	−0.021	0
	0.50	−0.106	−0.114	−0.132	−0.137	−0.139	−0.139	−0.101	−0.070	−0.043	−0.020	0
	0.60	−0.126	−0.130	−0.138	−0.141	−0.142	−0.142	−0.103	−0.070	−0.043	−0.020	0
	0.70	−0.137	−0.138	−0.142	−0.143	−0.143	−0.143	−0.104	−0.071	−0.043	−0.020	0
	0.80	−0.144	−0.144	−0.144	−0.144	−0.144	−0.144	−0.104	−0.071	−0.043	−0.020	0
$\dfrac{\bar{\tau}_{rz}}{\eta_1}$	0	0	0.433	1.025	1.450	$+\infty$	$+\infty$	1.093	0.820	0.658	0.527	0
	0.01	0	0.429	0.996	1.278	1.232	1.232	1.005	0.781	0.612	0.410	0
	0.02	0	0.418	0.920	1.024	0.966	0.966	0.846	0.683	0.507	0.267	0
	0.05	0	0.348	0.599	0.552	0.514	0.514	0.449	0.350	0.223	0.091	0
	0.10	0	0.176	0.188	0.141	0.122	0.122	0.095	0.058	0.022	−0.001	0
	0.15	0	0.031	−0.027	−0.045	−0.044	−0.044	−0.043	−0.044	−0.040	−0.027	0
	0.20	0	−0.052	−0.114	−0.110	−0.097	−0.097	−0.086	−0.072	−0.055	−0.031	0
	0.25	0	−0.084	−0.132	−0.118	−0.101	−0.101	−0.087	−0.071	−0.052	−0.028	0
	0.30	0	−0.086	−0.118	−0.102	−0.086	−0.086	−0.074	−0.059	−0.042	−0.022	0
	0.40	0	−0.056	−0.068	−0.056	−0.047	−0.047	−0.040	−0.031	−0.022	−0.011	0
	0.50	0	−0.027	−0.031	−0.025	−0.020	−0.020	−0.017	−0.013	−0.009	−0.005	0
	0.60	0	−0.011	−0.011	−0.009	−0.007	−0.007	−0.006	−0.005	−0.003	−0.001	0
	0.70	0	−0.003	−0.003	−0.002	−0.002	−0.002	−0.001	−0.001	−0.001	0.000	0
	0.80	0	0.000	0.000	0.000	0.000	0.000	0.000	0.000	0.000	0.000	0

$\gamma = 2; \mu = 0{,}8; \delta = 2$

	λ	Core						Shell				
	p	0	0.3	0.6	0.7	0.75	0.80	0.80	0.85	0.90	0.95	1.00
$\bar{\sigma}_z / \eta_1$	0	0	0	0	0	0	-0.016	0.464	0	0	0	0.6
	0.01	-0.084	-0.089	-0.118	-0.159	-0.212	-0.045	0.472	0.189	0.114	0.138	0.743
	0.02	-0.166	-0.176	-0.225	-0.267	-0.260	-0.056	0.487	0.300	0.238	0.346	0.862
	0.05	-0.397	-0.413	-0.432	-0.348	-0.222	-0.015	0.545	0.500	0.559	0.748	1.077
	0.10	-0.678	-0.669	-0.497	-0.284	-0.121	0.087	0.618	0.675	0.775	0.920	1.103
	0.15	-0.802	-0.749	-0.443	-0.215	-0.070	0.094	0.633	0.702	0.787	0.885	0.991
	0.20	-0.798	-0.714	-0.378	-0.185	-0.074	0.044	0.609	0.669	0.731	0.795	0.858
	0.25	-0.720	-0.629	-0.328	-0.182	-0.104	-0.026	0.576	0.619	0.662	0.702	0.739
	0.30	-0.616	-0.534	-0.295	-0.193	-0.142	-0.092	0.542	0.573	0.601	0.625	0.644
	0.40	-0.433	-0.385	-0.265	-0.223	-0.204	-0.188	0.495	0.507	0.517	0.524	0.528
	0.50	-0.326	-0.304	-0.257	-0.243	-0.238	-0.234	0.471	0.475	0.478	0.479	0.479
	0.60	-0.279	-0.271	-0.256	-0.253	-0.253	-0.253	0.462	0.463	0.463	0.463	0.462
	0.70	-0.262	-0.260	-0.257	-0.257	-0.257	-0.258	0.459	0.459	0.459	0.458	0.458
	0.80	-0.258	-0.258	-0.258	-0.258	-0.258	-0.258	0.459	0.459	0.459	0.459	0.459
$\bar{\sigma}_\theta / \eta_1$	0	-2.4	-2.4	-2.4	-2.4	-2.4	-2.215	-0.635	-0.6	-0.6	-0.6	-0.4
	0.01	-2.248	-2.238	-2.191	-2.142	-2.084	-1.914	-0.466	-0.461	-0.445	-0.400	-0.247
	0.02	-2.097	-2.078	-1.987	-1.900	-1.814	-1.673	-0.328	-0.328	-0.303	-0.244	-0.133
	0.05	-1.660	-1.617	-1.435	-1.302	-1.212	-1.108	-0.005	-0.006	0.017	0.058	0.113
	0.10	-1.037	-0.978	-0.772	-0.661	-0.598	-0.531	0.311	0.301	0.303	0.312	0.327
	0.15	-0.587	-0.537	-0.384	-0.313	-0.276	-0.238	0.466	0.446	0.433	0.425	0.420
	0.20	-0.304	-0.271	-0.178	-0.140	-0.121	-0.101	0.537	0.509	0.487	0.470	0.456
	0.25	-0.148	-0.130	-0.083	-0.066	-0.057	-0.049	0.564	0.532	0.505	0.483	0.464
	0.30	-0.076	-0.068	-0.049	-0.043	-0.040	-0.038	0.570	0.535	0.505	0.480	0.459
	0.40	-0.051	-0.052	-0.056	-0.058	-0.059	-0.061	0.558	0.521	0.490	0.463	0.440
	0.50	-0.075	-0.077	-0.083	-0.086	-0.087	-0.089	0.544	0.507	0.475	0.449	0.425
	0.60	-0.097	-0.098	-0.102	-0.104	-0.105	-0.105	0.536	0.499	0.467	0.441	0.418
	0.70	-0.109	-0.110	-0.112	-0.112	-0.113	-0.113	0.532	0.495	0.464	0.437	0.415
	0.80	-0.116	-0.116	-0.116	-0.116	-0.116	-0.116	0.530	0.493	0.462	0.436	0.414
$\bar{\sigma}_r / \eta_1$	0	-2.4	-2.4	-2.4	-2.4	-2.4	-1.276	-1.276	-0.6	-0.6	-0.6	0
	0.01	-2.248	-2.225	-2.096	-1.918	-1.661	-1.090	-1.090	-0.626	-0.429	-0.221	0
	0.02	-2.097	-2.053	-1.808	-1.522	-1.242	-0.907	-0.907	-0.558	-0.304	-0.089	0
	0.05	-1.660	-1.564	-1.128	-0.830	-0.665	-0.506	-0.506	-0.302	-0.138	-0.034	0
	0.10	-1.037	-0.914	-0.525	-0.366	-0.295	-0.236	-0.236	-0.140	-0.070	-0.024	0
	0.15	-0.587	-0.494	-0.266	-0.198	-0.172	-0.151	-0.151	-0.098	-0.055	-0.024	0
	0.20	-0.304	-0.253	-0.156	-0.135	-0.128	-0.124	-0.124	-0.084	-0.051	-0.024	0
	0.25	-0.148	-0.131	-0.110	-0.111	-0.112	-0.113	-0.113	-0.079	-0.050	-0.024	0
	0.30	-0.076	-0.077	-0.094	-0.103	-0.107	-0.111	-0.111	-0.077	-0.049	-0.023	0
	0.40	-0.051	-0.064	-0.094	-0.104	-0.108	-0.111	-0.111	-0.077	-0.048	-0.023	0
	0.50	-0.075	-0.084	-0.104	-0.110	-0.112	-0.114	-0.114	-0.078	-0.048	-0.023	0
	0.60	-0.097	-0.101	-0.111	-0.113	-0.114	-0.115	-0.115	-0.079	-0.048	-0.022	0
	0.70	-0.109	-0.111	-0.114	-0.115	-0.116	-0.116	-0.116	-0.079	-0.048	-0.022	0
	0.80	-0.116	-0.116	-0.116	-0.116	-0.116	-0.116	-0.116	-0.079	-0.049	-0.022	0
$\bar{\tau}_{rz} / \eta_1$	0	0	0.413	0.936	1.215	1.461	$+\infty$	$+\infty$	1.068	0.778	0.584	0
	0.01	0	0.410	0.917	1.157	1.287	1.234	1.234	0.973	0.720	0.463	0
	0.02	0	0.400	0.865	1.018	1.026	0.935	0.935	0.798	0.588	0.309	0
	0.05	0	0.337	0.602	0.578	0.521	0.454	0.454	0.367	0.241	0.100	0
	0.10	0	0.174	0.190	0.128	0.094	0.066	0.066	0.041	0.012	-0.007	0
	0.15	0	0.029	-0.040	-0.072	-0.078	-0.072	-0.072	-0.064	-0.053	-0.034	0
	0.20	0	-0.058	-0.134	-0.137	-0.127	-0.106	-0.106	-0.088	-0.066	-0.037	0
	0.25	0	-0.092	-0.151	-0.140	-0.124	-0.100	-0.100	-0.081	-0.059	-0.032	0
	0.30	0	-0.094	-0.134	-0.118	-0.102	-0.081	-0.081	-0.065	-0.046	-0.024	0
	0.40	0	-0.062	-0.076	-0.064	-0.054	-0.042	-0.042	-0.033	-0.023	-0.012	0
	0.50	0	-0.030	-0.034	-0.027	-0.022	-0.017	-0.017	-0.013	-0.009	-0.005	0
	0.60	0	-0.011	-0.012	-0.009	-0.007	-0.005	-0.005	-0.004	-0.003	-0.001	0
	0.70	0	-0.003	-0.003	-0.002	-0.002	-0.001	-0.001	-0.001	-0.001	0.000	0
	0.80	0	0.000	0.000	0.000	0.000	0.000	0.000	0.000	0.000	0.000	0

	λ	Core							Shell			
	p	0	0.3	0.6	0.7	0.75	0.80	0.85	0.85	0.90	0.95	1.00
$\frac{\bar{\sigma}_z}{\eta_1}$	0	0	0	0	0	0	0	-0.016	0.464	0	0	0.6
	0.01	-0.081	-0.086	-0.104	-0.124	-0.146	-0.191	0.004	0.487	0.216	0.191	0.808
	0.02	-0.161	-0.170	-0.202	-0.226	-0.240	-0.216	0.031	0.532	0.371	0.437	0.977
	0.05	-0.386	-0.399	-0.410	-0.354	-0.266	-0.095	0.192	0.661	0.691	0.877	1.230
	0.10	-0.661	-0.650	-0.489	-0.287	-0.125	0.090	0.360	0.765	0.875	1.029	1.222
	0.15	-0.782	-0.728	-0.424	-0.190	-0.037	0.140	0.340	0.759	0.855	0.963	1.076
	0.20	-0.773	-0.688	-0.342	-0.139	-0.020	0.108	0.240	0.712	0.781	0.851	0.918
	0.25	-0.689	-0.595	-0.280	-0.124	-0.040	0.045	0.130	0.656	0.703	0.746	0.784
	0.30	-0.578	-0.491	-0.238	-0.128	-0.073	-0.019	0.032	0.607	0.637	0.662	0.682
	0.40	-0.380	-0.328	-0.200	-0.154	-0.133	-0.074	-0.101	0.541	0.551	0.558	0.562
	0.50	-0.264	-0.241	-0.189	-0.174	-0.168	-0.145	-0.162	0.511	0.513	0.514	0.513
	0.60	-0.212	-0.204	-0.188	-0.185	-0.184	-0.176	-0.185	0.499	0.499	0.498	0.497
	0.70	-0.194	-0.192	-0.189	-0.189	-0.190	-0.188	-0.190	0.495	0.495	0.495	0.494
	0.80	-0.190	-0.190	-0.190	-0.190	-0.190	-0.190	-0.190	0.496	0.496	0.496	0.496
$\frac{\bar{\sigma}_\theta}{\eta_1}$	0	-2.4	-2.4	-2.4	-2.4	-2.4	-2.4	-2.215	-0.635	-0.6	-0.6	-0.4
	0.01	-2.247	-2.238	-2.197	-2.161	-2.128	-2.065	-1.890	-0.447	-0.431	-0.386	-0.228
	0.02	-2.096	-2.077	-1.997	-1.929	-1.872	-1.779	-1.625	-0.292	-0.275	-0.216	-0.099
	0.05	-1.657	-1.616	-1.446	-1.326	-1.242	-1.140	-1.020	0.054	0.070	0.110	0.167
	0.10	-1.026	-0.968	-0.764	-0.651	-0.586	-0.515	-0.440	0.365	0.364	0.371	0.385
	0.15	-0.567	-0.516	-0.359	-0.285	-0.245	-0.205	-0.164	0.508	0.492	0.482	0.476
	0.20	-0.276	-0.241	-0.145	-0.104	-0.083	-0.063	-0.043	0.570	0.545	0.525	0.509
	0.25	-0.115	-0.095	-0.046	-0.028	-0.019	-0.010	-0.002	0.590	0.561	0.535	0.514
	0.30	-0.040	-0.032	-0.012	-0.006	-0.004	-0.001	0.001	0.592	0.559	0.531	0.507
	0.40	-0.017	-0.018	-0.022	-0.025	-0.026	-0.028	-0.030	0.577	0.542	0.512	0.486
	0.50	-0.043	-0.045	-0.052	-0.055	-0.057	-0.058	-0.060	0.561	0.526	0.496	0.471
	0.60	-0.067	-0.069	-0.073	-0.075	-0.076	-0.076	-0.077	0.553	0.518	0.488	0.463
	0.70	-0.080	-0.081	-0.083	-0.084	-0.084	-0.085	-0.085	0.549	0.514	0.485	0.460
	0.80	-0.088	-0.088	-0.088	-0.088	-0.088	-0.088	-0.088	0.547	0.513	0.484	0.459
$\frac{\bar{\sigma}_r}{\eta_1}$	0	-2.4	-2.4	-2.4	-2.4	-2.4	-2.4	-1.276	-1.276	-0.6	-0.6	0
	0.01	-2.247	-2.226	-2.118	-2.002	-1.880	-1.610	-1.031	-1.031	-0.524	-0.240	0
	0.02	-2.096	-2.054	-1.846	-1.640	-1.451	-1.150	-0.792	-0.792	-0.403	-0.114	0
	0.05	-1.657	-1.565	-1.164	-0.882	-0.707	-0.526	-0.360	-0.360	-0.170	-0.044	0
	0.10	-1.026	-0.906	-0.513	-0.342	-0.264	-0.198	-0.150	-0.150	-0.077	-0.027	0
	0.15	-0.567	-0.472	-0.234	-0.161	-0.134	-0.114	-0.103	-0.103	-0.059	-0.026	0
	0.20	-0.276	-0.223	-0.119	-0.098	-0.093	-0.090	-0.088	-0.088	-0.055	-0.026	0
	0.25	-0.115	-0.096	-0.075	-0.077	-0.080	-0.083	-0.085	-0.085	-0.054	-0.026	0
	0.30	-0.040	-0.042	-0.061	-0.072	-0.078	-0.082	-0.085	-0.085	-0.054	-0.026	0
	0.40	-0.017	-0.030	-0.064	-0.076	-0.080	-0.084	-0.086	-0.086	-0.053	-0.025	0
	0.50	-0.043	-0.053	-0.075	-0.082	-0.084	-0.086	-0.087	-0.087	-0.054	-0.025	0
	0.60	-0.067	-0.072	-0.082	-0.085	-0.086	-0.087	-0.088	-0.088	-0.054	-0.025	0
	0.70	-0.080	-0.082	-0.086	-0.087	-0.088	-0.088	-0.088	-0.088	-0.054	-0.025	0
	0.80	-0.088	-0.088	-0.088	-0.088	-0.088	-0.088	-0.088	-0.088	-0.054	-0.025	0
$\frac{\bar{\tau}_{rz}}{\eta_1}$	0	0	0.399	0.877	1.093	1.238	1.463	$+\infty$	$+\infty$	1.007	0.677	0
	0.01	0	0.396	0.863	1.061	1.176	1.284	1.178	1.178	0.893	0.544	0
	0.02	0	0.387	0.823	0.974	1.028	1.009	0.870	0.870	0.683	0.364	0
	0.05	0	0.328	0.598	0.598	0.548	0.459	0.349	0.349	0.241	0.102	0
	0.10	0	0.172	0.191	0.121	0.076	0.032	0.004	0.004	-0.011	-0.018	0
	0.15	0	0.026	-0.053	-0.095	-0.108	-0.108	-0.091	-0.091	-0.071	-0.043	0
	0.20	0	-0.063	-0.153	-0.163	-0.156	-0.137	-0.104	-0.104	-0.077	-0.043	0
	0.25	0	-0.100	-0.170	-0.162	-0.147	-0.123	-0.089	-0.089	-0.065	-0.035	0
	0.30	0	-0.102	-0.149	-0.135	-0.119	-0.097	-0.068	-0.068	-0.049	-0.026	0
	0.40	0	-0.067	-0.085	-0.071	-0.061	-0.048	-0.032	-0.032	-0.023	-0.012	0
	0.50	0	-0.032	-0.037	-0.030	-0.025	-0.019	-0.013	-0.013	-0.009	-0.004	0
	0.60	0	-0.012	-0.013	-0.010	-0.008	-0.006	-0.004	-0.004	-0.002	-0.001	0
	0.70	0	-0.004	-0.003	-0.002	-0.002	-0.001	-0.001	-0.001	-0.001	0.000	0
	0.80	0	0.000	0.000	0.000	0.000	0.000	0.000	0.000	0.000	0.000	0

$\gamma = 2; \mu = 0,9; \delta = 2$

	λ	Core								Shell		
	p	0	0.3	0.6	0.7	0.75	0.80	0.85	0.90	0.90	0.95	1.00
$\dfrac{\bar{\sigma}_z}{\eta_1}$	0	0	0	0	0	0	0	0	−0.016	0.464	0	0.6
	0.01	−0.080	−0.083	−0.097	−0.104	−0.114	−0.129	−0.158	0.063	0.555	0.307	0.931
	0.02	−0.158	−0.165	−0.189	−0.197	−0.205	−0.201	−0.142	0.194	0.636	0.597	1.169
	0.05	−0.379	−0.389	−0.394	−0.348	−0.278	−0.142	0.115	0.553	0.854	1.040	1.419
	0.10	−0.649	−0.637	−0.481	−0.285	−0.127	0.089	0.372	0.723	0.952	1.120	1.336
	0.15	−0.766	−0.712	−0.407	−0.168	−0.011	0.175	0.385	0.616	0.901	1.021	1.142
	0.20	−0.752	−0.666	−0.311	−0.100	0.025	0.160	0.301	0.445	0.816	0.892	0.961
	0.25	−0.661	−0.564	−0.237	−0.072	0.016	0.106	0.196	0.282	0.735	0.780	0.819
	0.30	−0.542	−0.452	−0.187	−0.071	−0.011	0.046	0.101	0.150	0.669	0.694	0.715
	0.40	−0.330	−0.276	−0.139	−0.089	−0.068	−0.048	−0.032	−0.020	0.584	0.592	0.595
	0.50	−0.205	−0.180	−0.125	−0.108	−0.101	−0.096	−0.093	−0.091	0.548	0.548	0.548
	0.60	−0.149	−0.140	−0.122	−0.119	−0.118	−0.117	−0.118	−0.120	0.534	0.534	0.532
	0.70	−0.129	−0.126	−0.123	−0.123	−0.123	−0.124	−0.125	−0.125	0.531	0.530	0.530
	0.80	−0.125	−0.125	−0.125	−0.125	−0.125	−0.125	−0.125	−0.125	0.531	0.531	0.531
$\dfrac{\bar{\sigma}_\theta}{\eta_1}$	0	−2.4	−2.4	−2.4	−2.4	−2.4	−2.4	−2.4	−2.215	−0.635	−0.6	−0.4
	0.01	−2.246	−2.237	−2.198	−2.168	−2.143	−2.105	−2.034	−1.842	−0.410	−0.364	−0.196
	0.02	−2.093	−2.075	−2.000	−1.941	−1.895	−1.827	−1.720	−1.539	−0.228	−0.175	−0.048
	0.05	−1.650	−1.610	−1.447	−1.334	−1.256	−1.158	−1.037	−0.892	0.135	0.172	0.231
	0.10	−1.011	−0.954	−0.750	−0.635	−0.568	−0.494	−0.415	−0.333	0.426	0.433	0.445
	0.15	−0.544	−0.492	−0.331	−0.254	−0.213	−0.171	−0.128	−0.086	0.551	0.539	0.531
	0.20	−0.246	−0.210	−0.110	−0.068	−0.047	−0.026	−0.005	0.014	0.601	0.579	0.560
	0.25	−0.081	−0.061	−0.010	0.009	0.018	0.027	0.035	0.043	0.616	0.587	0.563
	0.30	−0.004	0.004	0.024	0.030	0.033	0.035	0.037	0.038	0.613	0.582	0.555
	0.40	0.018	0.017	0.012	0.009	0.007	0.005	0.003	0.001	0.595	0.562	0.533
	0.50	−0.011	−0.013	−0.020	−0.024	−0.025	−0.027	−0.029	−0.031	0.579	0.546	0.518
	0.60	−0.036	−0.038	−0.043	−0.044	−0.045	−0.046	−0.048	−0.049	0.570	0.537	0.510
	0.70	−0.051	−0.051	−0.054	−0.054	−0.055	−0.055	−0.056	−0.056	0.566	0.534	0.507
	0.80	−0.059	−0.059	−0.059	−0.059	−0.059	−0.059	−0.059	−0.059	0.565	0.533	0.505
$\dfrac{\bar{\sigma}_r}{\eta_1}$	0	−2.4	−2.4	−2.4	−2.4	−2.4	−2.4	−2.4	−1.276	−1.276	−0.6	0
	0.01	−2.246	−2.226	−2.128	−2.041	−1.957	−1.820	−1.525	−0.882	−0.882	−0.305	0
	0.02	−2.093	−2.053	−1.863	−1.700	−1.556	−1.341	−1.002	−0.590	−0.590	−0.170	0
	0.05	−1.650	−1.561	−1.180	−0.909	−0.738	−0.543	−0.350	−0.197	−0.197	−0.052	0
	0.10	−1.011	−0.891	−0.495	−0.315	−0.229	−0.156	−0.104	−0.078	−0.078	−0.029	0
	0.15	−0.544	−0.447	−0.201	−0.124	−0.094	−0.074	−0.063	−0.061	−0.061	−0.028	0
	0.20	−0.246	−0.191	−0.084	−0.061	−0.058	−0.056	−0.057	−0.058	−0.058	−0.028	0
	0.25	−0.081	−0.061	−0.041	−0.046	−0.049	−0.053	−0.057	−0.059	−0.059	−0.029	0
	0.30	−0.004	−0.007	−0.028	−0.041	−0.048	−0.054	−0.058	−0.059	−0.059	−0.028	0
	0.40	0.018	0.003	−0.033	−0.047	−0.051	−0.055	−0.058	−0.059	−0.059	−0.028	0
	0.50	−0.011	−0.021	−0.045	−0.053	−0.055	−0.058	−0.059	−0.059	−0.059	−0.028	0
	0.60	−0.036	−0.042	−0.053	−0.056	−0.058	−0.058	−0.059	−0.059	−0.059	−0.027	0
	0.70	−0.051	−0.053	−0.057	−0.058	−0.059	−0.059	−0.059	−0.059	−0.059	−0.027	0
	0.80	−0.059	−0.059	−0.059	−0.059	−0.059	−0.059	−0.059	−0.059	−0.059	−0.027	0
$\dfrac{\bar{\tau}_{rz}}{\eta_1}$	0	0	0.390	0.842	1.023	1.132	1.260	1.453	$+\infty$	$+\infty$	0.867	0
	0.01	0	0.387	0.830	1.001	1.096	1.191	1.262	1.074	1.074	0.680	0
	0.02	0	0.379	0.796	0.938	0.997	1.026	0.960	0.735	0.735	0.423	0
	0.05	0	0.322	0.593	0.608	0.570	0.486	0.352	0.195	0.195	0.083	0
	0.10	0	0.169	0.189	0.117	0.063	0.008	−0.035	−0.049	−0.049	−0.038	0
	0.15	0	0.022	−0.066	−0.118	−0.135	−0.142	−0.129	−0.089	−0.089	−0.053	0
	0.20	0	−0.070	−0.171	−0.187	−0.183	−0.166	−0.134	−0.084	−0.084	−0.047	0
	0.25	0	−0.108	−0.188	−0.183	−0.169	−0.146	−0.113	−0.068	−0.068	−0.036	0
	0.30	0	−0.110	−0.164	−0.152	−0.136	−0.114	−0.085	−0.049	−0.049	−0.026	0
	0.40	0	−0.072	−0.093	−0.080	−0.069	−0.056	−0.040	−0.022	−0.022	−0.012	0
	0.50	0	−0.035	−0.041	−0.033	−0.029	−0.023	−0.016	−0.009	−0.009	−0.004	0
	0.60	0	−0.014	−0.014	−0.011	−0.009	−0.007	−0.004	−0.001	−0.001	−0.001	0
	0.70	0	−0.004	−0.004	−0.003	−0.002	−0.002	−0.001	−0.001	−0.001	0.000	0
	0.80	0	0.000	0.000	0.000	0.000	0.000	0.000	0.000	0.000	0.000	0

	λ	Core									Shell	
	p	0	0.3	0.6	0.7	0.75	0.80	0.85	0.90	0.95	0.95	1.00
$\dfrac{\bar{\sigma}_z}{\eta_1}$	0	0	0	0	0	0	0	0	0	−0.016	0.464	0.6
	0.01	−0.079	−0.082	−0.094	−0.098	−0.098	−0.104	−0.105	−0.094	0.307	0.701	1.232
	0.02	−0.157	−0.163	−0.183	−0.187	−0.182	−0.178	−0.141	0.011	0.663	0.913	1.506
	0.05	−0.374	−0.384	−0.387	−0.342	−0.285	−0.164	0.062	0.484	1.186	1.191	1.618
	0.10	−0.641	−0.629	−0.475	−0.286	−0.127	0.085	0.372	0.734	1.160	1.176	1.401
	0.15	−0.753	−0.699	−0.393	−0.152	0.008	0.198	0.414	0.652	0.900	1.048	1.173
	0.20	−0.734	−0.646	−0.285	−0.069	0.059	0.199	0.345	0.494	0.640	0.916	0.988
	0.25	−0.636	−0.537	−0.200	−0.031	0.063	0.156	0.250	0.340	0.423	0.809	0.847
	0.30	−0.509	−0.417	−0.142	−0.021	0.040	0.102	0.160	0.213	0.258	0.726	0.746
	0.40	−0.285	−0.228	−0.084	−0.030	−0.006	0.016	0.035	0.050	0.059	0.626	0.630
	0.50	−0.151	−0.124	−0.064	−0.045	−0.037	−0.032	−0.027	−0.025	−0.026	0.584	0.583
	0.60	−0.089	−0.079	−0.059	−0.055	−0.053	−0.052	−0.053	−0.054	−0.056	0.569	0.568
	0.70	−0.067	−0.064	−0.059	−0.059	−0.060	−0.060	−0.061	−0.062	−0.063	0.565	0.564
	0.80	−0.061	−0.061	−0.061	−0.061	−0.061	−0.061	−0.061	−0.061	−0.061	0.566	0.566
$\dfrac{\bar{\sigma}_\theta}{\eta_1}$	0	−2.4	−2.4	−2.4	−2.4	−2.4	−2.4	−2.4	−2.4	−2.215	−0.635	−0.4
	0.01	−2.244	−2.236	−2.198	−2.169	−2.148	−2.116	−2.067	−1.973	−1.722	−0.319	−0.130
	0.02	−2.090	−2.072	−1.998	−1.943	−1.902	−1.844	−1.757	−1.613	−1.359	−0.105	0.034
	0.05	−1.642	−1.602	−1.442	−1.331	−1.255	−1.160	−1.039	−0.889	−0.710	0.240	0.302
	0.10	−0.995	−0.937	−0.732	−0.617	−0.548	−0.472	−0.390	−0.305	−0.219	0.488	0.500
	0.15	−0.520	−0.468	−0.304	−0.226	−0.184	−0.140	−0.096	−0.053	−0.012	0.591	0.581
	0.20	−0.216	−0.180	−0.078	−0.035	−0.014	0.008	0.029	0.049	0.067	0.631	0.610
	0.25	−0.048	−0.027	0.024	0.044	0.053	0.062	0.070	0.078	0.085	0.639	0.613
	0.30	0.030	0.039	0.059	0.065	0.068	0.070	0.072	0.074	0.075	0.634	0.605
	0.40	0.053	0.052	0.047	0.043	0.042	0.040	0.038	0.035	0.033	0.613	0.582
	0.50	0.023	0.020	0.013	0.009	0.007	0.005	0.004	0.002	0.000	0.597	0.566
	0.60	−0.004	−0.006	−0.011	−0.013	−0.014	−0.015	−0.016	−0.018	−0.019	0.588	0.557
	0.70	−0.020	−0.021	−0.023	−0.024	−0.024	−0.025	−0.026	−0.026	−0.027	0.584	0.554
	0.80	−0.030	−0.030	−0.030	−0.030	−0.030	−0.030	−0.030	−0.030	−0.030	0.582	0.552
$\dfrac{\bar{\sigma}_r}{\eta_1}$	0	−2.4	−2.4	−2.4	−2.4	−2.4	−2.4	−2.4	−2.4	−1.276	−1.276	0
	0.01	−2.244	−2.224	−2.130	−2.050	−1.988	−1.882	−1.709	−1.347	−0.573	−0.573	0
	0.02	−2.090	−2.050	−1.868	−1.717	−1.601	−1.419	−1.151	−0.733	−0.247	−0.247	0
	0.05	−1.642	−1.554	−1.179	−0.917	−0.740	−0.544	−0.328	−0.143	−0.046	−0.046	0
	0.10	−0.995	−0.875	−0.474	−0.286	−0.198	−0.116	−0.060	−0.031	−0.028	−0.028	0
	0.15	−0.520	−0.422	−0.170	−0.090	−0.059	−0.040	−0.029	−0.027	−0.029	−0.029	0
	0.20	−0.216	−0.161	−0.051	−0.030	−0.024	−0.025	−0.027	−0.030	−0.031	−0.031	0
	0.25	−0.048	−0.028	−0.008	−0.013	−0.020	−0.023	−0.028	−0.031	−0.031	−0.031	0
	0.30	0.030	0.028	0.005	−0.010	−0.016	−0.024	−0.029	−0.032	−0.031	−0.031	0
	0.40	0.053	0.038	−0.001	−0.016	−0.021	−0.026	−0.030	−0.031	−0.031	−0.031	0
	0.50	0.023	0.012	−0.014	−0.022	−0.026	−0.028	−0.030	−0.030	−0.030	−0.030	0
	0.60	−0.004	−0.010	−0.023	−0.026	−0.028	−0.029	−0.030	−0.030	−0.030	−0.030	0
	0.70	−0.020	−0.022	−0.027	−0.029	−0.029	−0.030	−0.030	−0.030	−0.030	−0.030	0
	0.80	−0.030	−0.030	−0.030	−0.030	−0.030	−0.030	−0.030	−0.030	−0.030	−0.030	0
$\dfrac{\bar{\tau}_{rz}}{\eta_1}$	0	0	0.386	0.826	0.996	1.083	1.186	1.292	1.432	$+\infty$	$+\infty$	0
	0.01	0	0.383	0.815	0.976	1.056	1.140	1.207	1.206	0.811	0.811	0
	0.02	0	0.374	0.782	0.920	0.980	1.017	1.002	0.838	0.390	0.390	0
	0.05	0	0.318	0.588	0.610	0.582	0.506	0.371	0.183	0.019	0.019	0
	0.10	0	0.165	0.183	0.107	0.052	−0.012	−0.066	−0.090	−0.056	−0.056	0
	0.15	0	0.017	−0.080	−0.137	−0.162	−0.171	−0.163	−0.127	−0.057	−0.057	0
	0.20	0	−0.076	−0.188	−0.208	−0.205	−0.193	−0.163	−0.115	−0.046	−0.046	0
	0.25	0	−0.115	−0.204	−0.202	−0.191	−0.168	−0.136	−0.092	−0.035	−0.035	0
	0.30	0	−0.117	−0.179	−0.168	−0.154	−0.132	−0.103	−0.068	−0.025	−0.025	0
	0.40	0	−0.078	−0.102	−0.090	−0.081	−0.066	−0.050	−0.032	−0.011	−0.011	0
	0.50	0	−0.038	−0.046	−0.039	−0.034	−0.027	−0.020	−0.012	−0.004	−0.004	0
	0.60	0	−0.015	−0.017	−0.014	−0.011	−0.009	−0.006	−0.004	−0.001	−0.001	0
	0.70	0	−0.005	−0.005	−0.004	−0.002	−0.002	−0.001	−0.001	0.000	0.000	0
	0.80	0	0.000	0.000	0.000	0.000	0.000	0.000	0.000	0.000	0.000	0